Statistical Methods for the Analysis of Genomic Data

Statistical Methods for the Analysis of Genomic Data

Special Issue Editors
Hui Jiang
Kevin He

MDPI • Basel • Beijing • Wuhan • Barcelona • Belgrade • Manchester • Tokyo • Cluj • Tianjin

Special Issue Editors
Hui Jiang
University of Michigan
USA

Kevin He
University of Michigan
USA

Editorial Office
MDPI
St. Alban-Anlage 66
4052 Basel, Switzerland

This is a reprint of articles from the Special Issue published online in the open access journal *Genes* (ISSN 2073-4425) (available at: https://www.mdpi.com/journal/genes/special_issues/statistical_methods).

For citation purposes, cite each article independently as indicated on the article page online and as indicated below:

LastName, A.A.; LastName, B.B.; LastName, C.C. Article Title. *Journal Name* **Year**, *Article Number*, Page Range.

ISBN 978-3-03936-140-3 (Hbk)
ISBN 978-3-03936-141-0 (PDF)

© 2020 by the authors. Articles in this book are Open Access and distributed under the Creative Commons Attribution (CC BY) license, which allows users to download, copy and build upon published articles, as long as the author and publisher are properly credited, which ensures maximum dissemination and a wider impact of our publications.

The book as a whole is distributed by MDPI under the terms and conditions of the Creative Commons license CC BY-NC-ND.

Contents

About the Special Issue Editors vii

Hui Jiang and Kevin He
Statistics in the Genomic Era
Reprinted from: *Genes* 2020, *11*, 443, doi:10.3390/genes11040443 1

Shuaichao Wang, Mengyun Wu and Shuangge Ma
Integrative Analysis of Cancer Omics Data for Prognosis Modeling
Reprinted from: *Genes* 2019, *10*, 604, doi:10.3390/genes10080604 5

Wanli Zhang and Yanming Di
Model-Based Clustering with Measurement or Estimation Errors
Reprinted from: *Genes* 2020, *11*, 185, doi:10.3390/genes11020185 25

Li Zeng, Zhaolong Yu and Hongyu Zhao
A Pathway-Based Kernel Boosting Method forSample Classification Using Genomic Data
Reprinted from: *Genes* 2019, *10*, 670, doi:10.3390/genes10090670 49

Qingyang Zhang
Testing Differential Gene Networks under Nonparanormal Graphical Models with False Discovery Rate Control
Reprinted from: *Genes* 2020, *11*, 167, doi:10.3390/genes11020167 63

Fengjiao Dunbar, Hongyan Xu, Duchwan Ryu, Santu Ghosh, Huidong Shi and Varghese George
Detection of Differentially Methylated Regions Using Bayes Factor for Ordinal Group Responses
Reprinted from: *Genes* 2019, *10*, 721, doi:10.3390/genes10090721 81

Mengli Xiao, Zhong Zhuang and Wei Pan
Local Epigenomic Data are more Informative than Local Genome Sequence Data in Predicting Enhancer-Promoter Interactions Using Neural Networks
Reprinted from: *Genes* 2020, *11*, 41, doi:10.3390/genes11010041 95

Fei Zhou, Jie Ren, Gengxin Li, Yu Jiang, Xiaoxi Li, Weiqun Wang and Cen Wu
Penalized Variable Selection for Lipid–Environment Interactions in a Longitudinal Lipidomics Study
Reprinted from: *Genes* 2019, *10*, 1002, doi:10.3390/genes10121002 111

About the Special Issue Editors

Hui Jiang is Associate Professor of Biostatistics at the University of Michigan. He received his Ph.D. in Computational and Mathematical Engineering from Stanford University in 2009. He is interested in developing statistical and computational methods for the analysis of large-scale biological data generated using modern high-throughput technologies.

Kevin He is Research Assistant Professor of Biostatistics at the University of Michigan. He received his Ph.D. in Biostatistics from the University of Michigan in 2012. His research interests include survival analysis, high-dimensional data analysis, statistical genetics and genomics, and statistical methods for epidemiology, in addition to the development of statistical optimization methods for analyzing large-scale databases

Editorial

Statistics in the Genomic Era

Hui Jiang * and Kevin He

Department of Biostatistics, University of Michigan, Ann Arbor, MI 48109, USA; kevinhe@umich.edu
* Correspondence: jianghui@umich.edu

Received: 14 April 2020; Accepted: 15 April 2020; Published: 18 April 2020

In recent years, technology breakthroughs have greatly enhanced our ability to understand the complex world of molecular biology. Rapid developments in genomic profiling techniques, such as high-throughput sequencing, have brought new opportunities and challenges to the fields of computational biology and bioinformatics. Furthermore, by combining genomic profiling techniques with other experimental techniques, many powerful approaches (e.g., RNA-Seq, Chips-Seq, single-cell assays, and Hi-C) have been developed in order to help explore the complex biological systems. As more genomic datasets become available, both in volume and variety, the analysis of such data has become a critical challenge as well as a topic of interest. Consequently, statistical methods dealing with the problems associated with these newly developed techniques are in high demand. This special issue of *Genes*, titled *Statistical Methods for the Analysis of Genomic Data*, consists of a number of studies which highlight the state-of-the-art statistical methods for the analysis of genomic data and explore future directions for improvement.

Gene expression is one of the most widely studied topics in genomics. From microarray [1] to high-throughput sequencing of transcriptomes (RNA-Seq) [2], expression levels of tens of thousands of genes can be measured simultaneously. After such data are collected, the first analysis is often to identify genes whose expression levels are associated with experimental conditions or outcomes. Depending on the type of variables, the initial analysis can be done using two-group comparisons (a.k.a. differential expression), linear or Cox regressions, or more complicated statistical models. In clinical studies, the statistical power to identify biologically relevant genes is often limited by the scarce patient samples, which is especially the case for rare diseases such as cancers. Integrated analysis can help improve statistical power by borrowing information across multiple datasets. In [3], Wang et al., introduce a novel penalized regression-based approach for the integrated analysis of gene expression data with survival outcomes. Novel shrinkage penalty functions are proposed to promote similarity among estimated coefficients from each cancer, and the coordinate descent (CD) algorithm is used for model fitting. The proposed method is applied to gene expression data measured using RNA-Seq from The Cancer Genome Atlas (TCGA) project [4] on nine different cancers, and identifies potentially informative genes that are prognostic for patient survival times in multiple cancers.

Due to the large number of genes in a typical genome (e.g., ~25,000 protein coding genes in the human genome), the initial differential expression analysis often identifies many potentially informative genes. To further understand the underlying biology, unsupervised clustering analysis is often conducted to group genes with similar expression patterns together. In the current standard practice, the estimation errors in the gene fold-changes during the initial differential expression analysis are often ignored in the downstream clustering analysis. To address this problem, in [5], Zhang and Di present a novel clustering approach, named MCLUST-ME, which takes the estimation errors in the gene fold-changes into consideration. The proposed model combines the conventional Gaussian mixture clustering model in MCLUST [6] with a random Gaussian measurement error assuming a known variance for each observation, and uses an extended Expectation–Maximization (EM) algorithm for model fitting. A unique feature of MCLUST-ME is that the classification boundary depends on the distribution of the measurement error for each observation, which is shown to achieve improved clustering performance in an RNA-Seq dataset on *Arabidopsis thaliana*.

The analysis of cancer genomic data has long suffered the curse of dimensionality, as sample sizes for most cancer genomic studies are a few hundred at most, while tens of thousands of genomic features are studied. To leverage prior biological knowledge, such as pathways, and more effectively analyze cancer genomic data, the research article by Zeng et al., [7] proposes a Pathway-based Kernel Boosting (PKB) method for integrating gene pathway information for sample classification; the authors use kernel functions calculated from each pathway as base learners and learn the weights through an iterative optimization of the classification loss function. Instead of the first-order approximation used in the usual gradient descent boosting method, used by Wei and Li [8] and Luan and Li [9], the PKB approach uses the second-order approximation of the loss function, which allows for deeper descent at each step. Moreover, the PKB includes two types of regularizations (L1 and L2) for the selection of base learners in each iteration and outperforms other methods, identifying pathways relevant to the outcome variables. The proposed method is applied to gene expression datasets on three cancer types, including breast cancer, melanoma, and glioma, and outperforms competing methods in terms of the prediction of clinical features including tumor grade, tumor site, and metastasis status, as well as the identification of relevant gene pathways.

To study the different roles of the cell cycle pathway in the two subtypes of breast cancer, including luminal A subtype and basal-like subtype using a TCGA (The Cancer Genome Atlas) gene expression dataset, Zhang [10] considers a computational pipeline of detecting differential substructure between two nonparanormal graphical models with false discovery rate control. The proposed approach extends the hierarchical testing method introduced by Liu [11] to a more flexible semiparametric framework and provides a convenient tool for modeling the dependency structure between non-Gaussian data while maintaining the good interpretability and computational convenience of Gaussian graphical models.

Besides transcriptomics, epigenomics has also undergone rapid development in recent years, which provides complementary information for studying cellular functions on top of transcriptomics. Detecting differentially methylated regions (DMRs) based on reduced representation bisulfite sequencing (RRBS) has been widely employed for identifying regions in the genome where the methylation status is associated with the phenotype of interest [12]. Till now, existing methods have been mostly focused on binary phenotypes. Dunbar et al. [13] developed a novel Bayes Factor Method (BFM) to detect genomic loci that are associated with ordinal group responses. Mixed-effect modeling is used to accommodate the correlated methylation states among neighboring CpG (5'—C—phosphate—G—3') sites. The proposed method is applied to bisulfite sequencing data from a chronic lymphocytic leukemia (CLL) study.

Enhancer-promoter interactions (EPIs) give important information for understanding transcriptional regulation inside cells. However, experimentational approaches investigating EPIs, such as Hi-C [14], are laborious and expensive. Recently, using existing genomic data and machine learning methods to predict EPIs has shown promising results. Xiao et al. [15] have conducted a rigorous study comparing various machine learning methods including convolutional neural networks (CNNs), feed-forward neural networks (FNNs), and gradient boosting with local sequence and 22 epigenomic data types from the K562 cell line on their predictive powers for Epos By randomly splitting the chromosomes rather than the enhancer-promoter pairs, duplication and overlapping cases between training and testing sets are avoided. As a result, they found that local epigenomic features are more predictive of EPIs than local sequences, and combining the two does not provide much predictive gain.

Last but not least, Zhou et al. [16] has developed a novel penalized variable selection method to identify important lipid—environment interactions in a longitudinal lipidomics study. Lipid species play key roles in many biological processes such as signal transduction, cell homeostasis, and energy storage. The authors propose an efficient Newton-Raphson-based algorithm within the generalized estimating equation (GEE) framework. Compared with existing penalization methods [17–20] in longitudinal studies that have been mostly developed for the identification of important main effects only, the proposed procedure simultaneously selects individual main effect and group structure corresponding to the main lipid effect and interaction effect respectively. The proposed method is applied to a high-dimensional longitudinal lipid dataset from 60 female CD-1 mice in four different treatment groups and identifies markers that show potential association with body weight.

Biologists and statisticians do not always speak the same language, but when they do, the interplay and synergy between them can dramatically advance science. In the modern genomic era, we hope this special issue showcases in a timely manner how novel statistical methods can help improve genomic data analysis, and vice versa, how new challenges in genomic data analysis can inspire method development in statistics.

Author Contributions: Writing, H.J. and K.H. All authors have read and agreed to the published version of the manuscript.

Funding: This research received no external funding.

Conflicts of Interest: The authors declare no conflict of interest.

References

1. Schena, M.; Shalon, D.; Heller, R.; Chai, A.; Brown, P.O.; Davis, R.W. Parallel human genome analysis: Microarray-based expression monitoring of 1000 genes. *Proc. Natl. Acad. Sci. USA* **1996**, *93*, 10614–10619. [CrossRef] [PubMed]
2. Mortazavi, A.; Williams, B.A.; McCue, K.; Schaeffer, L.; Wold, B. Mapping and quantifying mammalian transcriptomes by RNA-Seq. *Nat. Methods* **2008**, *5*, 621–628. [CrossRef] [PubMed]
3. Wang, S.; Wu, M.; Ma, S. Integrative Analysis of Cancer Omics Data for Prognosis Modeling. *Genes* **2019**, *10*, 604. [CrossRef] [PubMed]
4. Tomczak, K.; Czerwińska, P.; Wiznerowicz, M. The Cancer Genome Atlas (TCGA): An immeasurable source of knowledge. *Contemp. Oncol.* **2015**, *19*, A68. [CrossRef] [PubMed]
5. Zhang, W.; Di, Y. Model-Based Clustering with Measurement or Estimation Errors. *Genes* **2020**, *11*, 185. [CrossRef] [PubMed]
6. Fraley, C.; Raftery, A.E. Enhanced model-based clustering, density estimation, and discriminant analysis software: MCLUST. *J. Classif.* **2003**, *20*, 263–286. [CrossRef]
7. Zeng, L.; Yu, Z.; Zhao, H. A Pathway-Based Kernel Boosting Method for Sample Classification Using Genomic Data. *Genes* **2019**, *10*, 670. [CrossRef] [PubMed]
8. Wei, Z.; Li, H. Nonparametric pathway-based regression models for analysis of genomic data. *Biostatistics* **2007**, *8*, 265–284. [CrossRef] [PubMed]
9. Luan, Y.; Li, H. Group additive regression models for genomic data analysis. *Biostatistics* **2008**, *9*, 100–113. [CrossRef] [PubMed]
10. Zhang, Q. Testing Differential Gene Networks under Nonparanormal Graphical Models with False Discovery Rate Control. *Genes* **2020**, *11*, 167. [CrossRef] [PubMed]
11. Liu, W. Structural similarity and difference testing on multiple sparse Gaussian graphical models. *Ann. Stat.* **2017**, *45*, 2680–2707. [CrossRef]
12. Meissner, A.; Gnirke, A.; Bell, G.W.; Ramsahoye, B.; Lander, E.S.; Jaenisch, R. Reduced representation bisulfite sequencing for comparative high-resolution DNA methylation analysis. *Nucleic Acids Res.* **2005**, *33*, 5868–5877. [CrossRef] [PubMed]
13. Dunbar, F.; Xu, H.; Ryu, D.; Ghosh, S.; Shi, H.; George, V. Detection of Differentially Methylated Regions Using Bayes Factor for Ordinal Group Responses. *Genes* **2019**, *10*, 721. [CrossRef] [PubMed]
14. Belton, J.M.; McCord, R.P.; Gibcus, J.H.; Naumova, N.; Zhan, Y.; Dekker, J. Hi–C: A comprehensive technique to capture the conformation of genomes. *Methods* **2012**, *58*, 268–276. [CrossRef] [PubMed]
15. Xiao, M.; Zhuang, Z.; Pan, W. Local Epigenomic Data are more Informative than Local Genome Sequence Data in Predicting Enhancer-Promoter Interactions Using Neural Networks. *Genes* **2020**, *11*, 41. [CrossRef] [PubMed]
16. Zhou, F.; Ren, J.; Li, G.; Jiang, Y.; Li, X.; Wang, W.; Wu, C. Penalized Variable Selection for Lipid–Environment Interactions in a Longitudinal Lipidomics Study. *Genes* **2019**, *10*, 1002. [CrossRef] [PubMed]
17. Wang, L.; Zhou, J.; Qu, A. Penalized Generalized Estimating Equations for High-Dimensional Longitudinal Data Analysis. *Biometrics* **2012**, *68*, 353–360. [CrossRef] [PubMed]

18. Ma, S.; Song, Q.; Wang, L. Simultaneous variable selection and estimation in semiparametric modeling of longitudinal/clustered data. *Bernoulli* **2013**, *19*, 252–274. [CrossRef]
19. Cho, H.; Qu, A. Model selection for correlated data with diverging number of parameters. *Stat. Sin.* **2013**, *23*, 901–927. [CrossRef]
20. Fan, Y.; Qin, G.; Zhu, Z. Variable selection in robust regression models for longitudinal data. *J. Multivar. Anal.* **2012**, *109*, 156–167. [CrossRef]

© 2020 by the authors. Licensee MDPI, Basel, Switzerland. This article is an open access article distributed under the terms and conditions of the Creative Commons Attribution (CC BY) license (http://creativecommons.org/licenses/by/4.0/).

Article

Integrative Analysis of Cancer Omics Data for Prognosis Modeling

Shuaichao Wang [1], Mengyun Wu [2,*] and Shuangge Ma [3,*]

1. School of Life Sciences and Biotechnology, Shanghai Jiao Tong University, Shanghai 200240, China
2. School of Statistics and Management, Shanghai University of Finance and Economics, Shanghai 200433, China
3. Department of Biostatistics, Yale University, New Haven, CT 06520, USA
* Correspondence: wu.mengyun@mail.shufe.edu.cn (M.W.); shuangge.ma@yale.edu (S.M.)

Received: 13 July 2019; Accepted: 7 August 2019; Published: 9 August 2019

Abstract: Prognosis modeling plays an important role in cancer studies. With the development of omics profiling, extensive research has been conducted to search for prognostic markers for various cancer types. However, many of the existing studies share a common limitation by only focusing on a single cancer type and suffering from a lack of sufficient information. With potential molecular similarity across cancer types, one cancer type may contain information useful for the analysis of other types. The integration of multiple cancer types may facilitate information borrowing so as to more comprehensively and more accurately describe prognosis. In this study, we conduct marginal and joint integrative analysis of multiple cancer types, effectively introducing integration in the discovery process. For accommodating high dimensionality and identifying relevant markers, we adopt the advanced penalization technique which has a solid statistical ground. Gene expression data on nine cancer types from The Cancer Genome Atlas (TCGA) are analyzed, leading to biologically sensible findings that are different from the alternatives. Overall, this study provides a novel venue for cancer prognosis modeling by integrating multiple cancer types.

Keywords: multiple cancer types; integrative analysis; omics data; prognosis modeling

1. Introduction

Cancer is one of the leading causes of death worldwide and has been posing extensive public concerns. In cancer studies, prognosis modeling is a critical step that greatly contributes to understanding cancer etiology, developing effective therapeutic methods, and improving life quality. Significant effort has been devoted to searching for prognostic factors, among which omics markers have important implications. For example, *EGFR* has been suggested as a strong prognostic indicator in multiple cancers, such as ovarian, cervical, and bladder cancers. Nicholson, et al. [1] reviewed over 200 studies and reported that relapse-free-interval or survival data are directly in relation to the increased EGFR levels in breast, gastric, colorectal, and many other cancers. Petitjean, et al. [2] found that the mutation of *TP53* has an impact on the prognosis of breast and several other cancers. Gao, et al. [3] used a Cox model to find that a high level of MMP-14 mRNA expression leads to a significantly shorter overall survival for breast cancer. Chiu, et al. [4] characterized prognostic alteration for melanoma with a panel of five genes, including *CSMD2, CNTNAP5, NRDE2, ADAM6,* and *TRPM2*. Despite considerable successes, our understanding of cancer prognosis is still limited. The limited progress in cancer analytics may be attributable to small sample sizes, high dimensionality and low signal-to-noise ratios of omics data, as well as the underlying molecular complexity of cancers.

Most of the existing studies, including the aforementioned, focus on a single type of cancer, and analysis often suffers from a lack of sufficient information. Cancer types have been typically classified according to organ- and tissue histology-based pathology criteria. This is especially true in "old" studies.

More recently, with the development of high-throughput profiling, increasing attention has been paid to the molecular basis of cancers, providing a novel perspective on cancer types. A representative recent work is Hoadley, et al. [5], which conducted the molecular clustering of 33 different types of tumors in The Cancer Genome Atlas (TCGA) with data on aneuploidy, DNA methylation, mRNA, and miRNA. Their results show that some cancers, which were treated as completely different diseases according to traditional organ- and tissue histology-based pathology criteria, are closely related according to their molecular characteristics. For example, squamous cell carcinoma can occur in lung, bladder, cervix, head, and neck, and different histopathological types are often observed. However, in Hoadley, et al. [5], these cancer types have been found to have similar molecular characteristics.

Molecular similarity across cancers has been well established in the literature. Prognosis of many different cancer types is mediated by some common mechanisms associated with certain common pathways. For example, the p53 pathway inhibits cell growth and stimulates cell death, which plays an important role in a large fraction of cancers. In addition, there are other genes/pathways that have important roles in many cancer types, such as apoptosis, hypoxia-inducible transcription factor (HIF)-1, mitogen activated protein kinase (MAPK) phosphoinositide3-kinase (PI3K), and receptor tyrosine kinases (RTKs) [6]. Published studies have found that different cancer types may share common oncogenes, tumor-suppressor genes and stability genes, the alternations of which are responsible for the genesis and prognosis of cancers. For example, *BRCA1* gene mutation is often found in both breast and ovarian cancers [7]. These two cancer types are perhaps the most common cancers in female and often occur together [7]. Another example is lung adenocarcinoma and lung squamous cell carcinoma which are two major lung cancer subtypes. Many genes have been reported to be associated with both cancer subtypes, including *EGFR* [8], *TP53* [8], *AKT1*, *DDR2* [9], *FGFR1* [10], *KRAS* [8], *PTEN*, and others. With molecular similarity, one cancer may contain information useful for the analysis of other cancers. Overall, it is of interest and also reasonable to conduct the integrative analysis of molecular profiles of multiple cancer types to increase information and more accurately describe the underlying prognosis.

More recently, much effort has been devoted to collecting omics profiles of tumor samples with different cancer types under a unified protocol. A representative example is TCGA organized by The National Cancer Institute (NCI) which has generated a large amount of cross-platform genomic data for exploring the complex landscapes of human cancers. Specifically, it has collected multi-omics data from over 20,000 primary cancer and matched normal samples spanning 33 cancer types, including breast cancer, lung squamous cell carcinoma, lung adenocarcinoma, and others. Other examples include the International Cancer Genome Consortium (ICGC), Therapeutically Applicable Research to Generate Effective Treatments (TARGET), and others. With the clinical and omics data on multiple cancer types, these databases provide a good opportunity to conduct cancer modeling through data integration.

In the literature, there are a few related studies, which can be generally classified into two families. The first family adopts a meta-analysis strategy, which first analyzes different cancer types separately and then compares results across cancer types to search for overlapping findings. An example is Cava, et al. [11], which first analyzed gene expression data on 16 cancer types separately and then identified 895 de-regulated genes with a central role in pathways. Yu, et al. [12] systematically analyzed gene expressions across diverse cancers during the inflammatory timeline. After comparing the differentially expressed genes among cancers, they found three novel pan-cancer gene expression patterns, in which the gene expressions are regulated differently in the early and late phases of inflammation. Using a cohort of 3899 samples with 10 cancer types, Sharma, et al. [13] adopted a bottom-up approach to quantify the effects of gene expression variations and identified novel recurrent regulatory mutations influencing known cancer genes, such as *GRIN2D* and *NKX2-1*, in multiple cancer types. The second family of approaches stacks data from multiple cancer types together to create a "mega" dataset, and then conducts analysis as if there is in fact just a single dataset. An example is Martinez-Ledesma, et al. [14], which used a network-based exploration approach to identify gene expression biomarkers that are predictive of clinical outcomes in 12 cancer types. Using TCGA data on

3281 samples with 12 cancer types, Leiserson, et al. [15] performed a pan-cancer analysis of mutated networks with a new algorithm, HotNet2, and found some significantly mutated subnetworks as well as those with less characterized roles in cancers. Beyond studies on cancer omics data, similar strategies have also been considered in other fields of biomedical research to collectively analyze multiple datasets. For example, Xing, et al. [16] proposed two variations of a stacking algorithm to simultaneously predict the resistance of multiple drugs using mutation information, leading to improvement in prediction performance. As another example of drug analysis, Matlock, et al. [17] developed stacking models built on multiple cell lines, multiple tested drugs, as well as genomic information for drug sensitivity prediction in cancer cell lines. Medical imaging data integration has also been conducted. For example, a meta-analysis based support vector machine was introduced in [18] to collectively analyze multiple types of images, such as fluorodeoxyglucose positron emission tomography (FDG-PET) and magnetic resonance imaging (MRI), for identifying susceptible brain regions and predicting the incidence of Alzheimer's disease.

Despite considerable successes, both families have limitations. The former neglects integration in the discovery process. Data on each cancer type still suffers from a lack of sufficient information resulting from a small sample size, high noises, and other reasons. As such, the "delay" in integration may make the analysis less effective. For the latter one, although sample size increases by stacking, subjects with different cancer types are treated as if they were from the same population. It cannot effectively accommodate the heterogeneity across cancer types. In addition, in some of the existing studies, "classic" statistical techniques have been adopted, and there is a lack of utilizing state-of-the-art techniques.

Motivated by the limitations of single cancer type analysis and recent successes of integrative analysis in other contexts, in this study our goal is to conduct more effective integrative analysis of multiple cancer types with high dimensional omics data. By contrast with the single cancer type analysis, omics data from multiple cancer types are jointly analyzed to effectively borrow information across cancer types and generate more reliable findings. By contrast with the existing meta-analysis- and stacking-based approaches, the proposed analysis integrates data on multiple cancer types in the discovery process and effectively accommodate the heterogeneity across cancer types. By contrast with the analysis on categorical and continuous outcomes, the more challenging prognosis analysis is conducted. The proposed analysis is based on the penalization technique which has a solid statistical ground and satisfactory performance in published studies. TCGA mRNA expression data on nine cancer types are analyzed to demonstrate the proposed integrative analysis approach. Overall, this study provides a practically useful new venue for cancer prognosis modeling with multiple cancer types.

2. Materials and Methods

2.1. The Cancer Genome Atlas (TCGA) Data

TCGA is one of the largest cancer genomics programs that comprehensively cover multiple cancer types with high quality omics measurements and serves as an ideal testbed. In this study, the processed level 3 data are downloaded from cBioPortal (http://www.cbioportal.org/). For omics data, we consider mRNA expressions which were measured using the IlluminaHiseq RNAseq V2 platform. For each subject, a total of 20,531 mRNA expression measurements are available. It is noted that the proposed analysis can be directly applied to other types of omics data, such as copy number variation, methylation, microRNA, and others. The prognosis outcome of interest is the overall survival time which is subject to right censoring. Nine common cancer types are analyzed, including some recognized as highly correlated, such as lung adenocarcinoma and lung squamous cell carcinoma. Summary information is provided in Table 1. We acknowledge that, as the proposed analysis can well accommodate heterogeneity across cancers, the selection of cancers for analysis does not need to follow a strict criterion. Beyond these nine cancers with high prevalence and mortality, others can be added to the analysis easily.

Table 1. Summary information of the nine cancer types.

Cancer Type	Abbreviation	Sample Size	Non-Censored	Overall Survival (Month)	Median Survival
Breast invasive carcinoma	BRCA	802	119	0.03–282.69	29.88
Bladder Urothelial Carcinoma	BLCA	409	180	0.43–165.90	17.61
Glioblastoma multiforme	GBM	541	417	0.10–127.60	10.70
Head and Neck squamous cell carcinoma	HNSC	159	69	0.07–135.19	12.48
Acute Myeloid Leukemia	LAML	199	132	0.10–118.10	17.00
Lung adenocarcinoma	LUAD	509	183	0.13–238.11	21.62
Lung squamous cell carcinoma	LUSC	497	215	0.03–173.69	21.91
Ovarian serous cystadenocarcinoma	OV	582	384	0.26–180.06	33.03
Pancreatic adenocarcinoma	PAAD	184	100	0.13–90.05	15.34

It has been suggested in the literature that the number of important prognostic markers is not expected to be large. Besides, with a relatively moderate sample size for each cancer type and a much larger number of genes, analysis may not be reliable. To improve estimation stability and also reduce computational cost, we conduct prescreening as follows. We consider the 1385 genes in the TruSight RNA Pan-Cancer Panel which is produced by Illunima Company and provides a comprehensive assessment of cancer-related RNA transcripts and fusion detection. These genes have been referred to in public databases and implicated in multiple cancer types, including solid tumors, soft tissue cancers, and hematological malignancies [19]. After data matching, a total of 1040 gene expression measurements are left for downstream analysis. Note that this prescreening is not essential in our analysis, and the proposed approach can be directly applied to a bigger set of genes.

2.2. Methods

We conduct both marginal and joint analysis, where the former analyzes one gene at a time and the latter analyzes all genes in a single model. Both types of analysis have been extensively conducted in existing cancer modeling studies. As they have different implications and cannot replace each other, we conduct both analyses to generate a more comprehensive understanding of cancer prognosis. We develop a penalized regression-based framework to collectively analyze multiple datasets and identify markers associated with the prognosis of multiple cancer types, while effectively accounting for the similarity across cancers. The overall flowchart of analysis is provided in Figure 1.

Assume that there are K cancer types, where the kth ($k = 1, \ldots, K$) type has $n^{(k)}$ independent subjects. For subject i with the kth cancer type, let $T_i^{(k)}$ be the log-transformed survival time and $X_i^{(k)} = \left(X_{i1}^{(k)}, \ldots, X_{ip}^{(k)}\right)$ be the p-dimensional vector of gene expression measurements. In practical analysis, right censoring is usually present. Denote $C_i^{(k)}$ as the log-transformed censoring time, then we observe $y_i^{(k)} = \min\left(T_i^{(k)}, C_i^{(k)}\right)$ and $\delta_i^{(k)} = I\left(T_i^{(k)} \leq C_i^{(k)}\right)$ with $I(\cdot)$ being the indicator function.

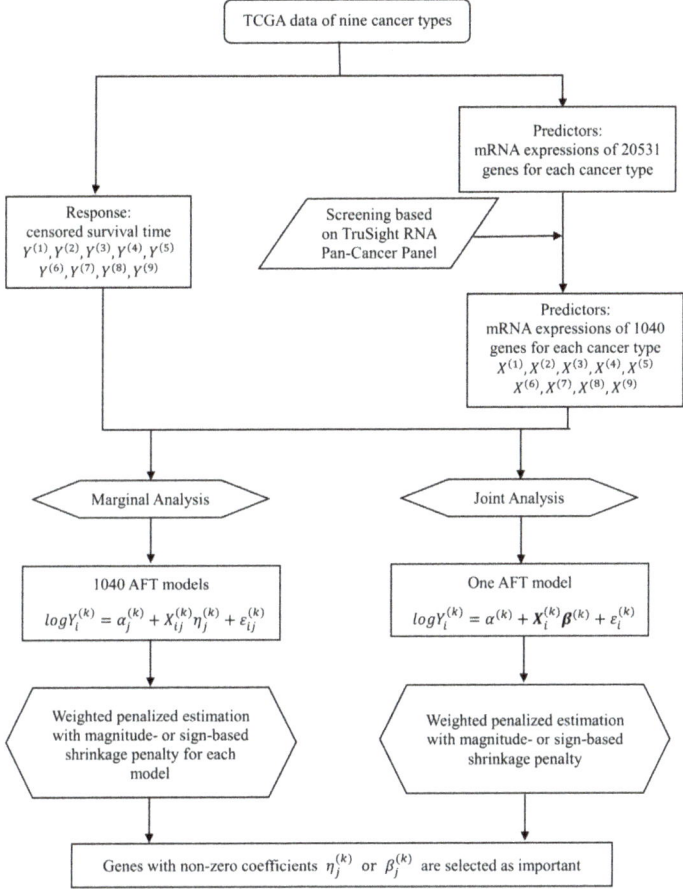

Figure 1. Flowchart of the proposed integrative analysis of The Cancer Genome Atlas (TCGA) data.

2.2.1. Marginal Analysis

We adopt the accelerated failure time (AFT) model for describing prognosis. It has been one of the most popular choices in high-dimensional survival analysis due to its lucid interpretation and, more importantly, computational simplicity [20]. For a specific cancer type, consider the marginal AFT model for the jth measurement as:

$$T_i^{(k)} = \alpha_j^{(k)} + X_{ij}^{(k)} \eta_j^{(k)} + \varepsilon_{ij}^{(k)}, \tag{1}$$

where $\alpha_j^{(k)}$ and $\eta_j^{(k)}$ are the unknown intercept and coefficient, and $\varepsilon_{ij}^{(k)}$ is the random error. Assume that for each cancer type, data $\{\{X_i^{(k)}, y_i^{(k)}, \delta_i^{(k)}\}, i = 1, \ldots, n^{(k)}\}$ have been sorted according to $y_i^{(k)}$ in an ascending order. Then, the following weighted penalized objective function is proposed to collectively analyze multiple cancer types,

$$\sum_{k=1}^{K} \left[\frac{1}{2n^{[k]}} \sum_i w_i^{[k]} \left[y_i^{[k]} - \alpha_j^{[k]} - x_{ij}^{[k]} \eta_j^{[k]} \right]^2 \right] + \sum_{k=1}^{K} \rho_{MCP}\left(\eta_j^{(k)}, \lambda_1, \gamma\right) + \frac{\lambda_2}{2} \sum_{k' \neq k}^{K} \rho\left(\eta_j^{(k)}, \eta_j^{(k')}\right) \tag{2}$$

Here, $w_i^{(k)}$'s are the Kaplan–Meier (KM) weights for accommodating censoring and defined as

$$w_1^{(k)} = \frac{\delta_1^{(k)}}{n^{(k)}}, w_i^{(k)} = \frac{\delta_i^{(k)}}{n^{(k)} - i + 1} \prod_{l=1}^{i-1} \left(\frac{n^{(k)} - l}{n^{(k)} - l + 1}\right)^{\delta_i^{(k)}}, \ i = 2, \ldots, n^{(k)}$$

$\rho_{MCP}(|v|, \lambda_1, \gamma) = \lambda_1 \int_0^{|v|} \left(1 - \frac{x}{\lambda_1 \gamma}\right)_+ dx$ is the minimax concave penalty (MCP) with tuning parameter λ_1 and regularization parameter γ. We consider two types of $\rho\left(\eta_j^{(k)}, \eta_j^{(k')}\right)$ with tuning parameter λ_2. The first is the magnitude-based shrinkage penalty with

$$\rho\left(\eta_j^{(k)}, \eta_j^{(k')}\right) = \left(\eta_j^{(k)} - s_j^{(kk')} \eta_j^{(k')}\right)^2, \quad (3)$$

where $s_j^{(kk')} = I\left(\text{Sgn}\left(\eta_j^{(k)}\right) = \text{Sgn}\left(\eta_j^{(k')}\right)\right)$ with $\text{Sgn}(\cdot)$ being the sign function. The second is the sign-based shrinkage penalty with

$$\rho\left(\eta_j^{(k)}, \eta_j^{(k')}\right) = \left(\text{Sgn}\left(\eta_j^{(k)}\right) - \text{Sgn}\left(\eta_j^{(k')}\right)\right)^2 \quad (4)$$

Based on (2), a total of p objective functions are developed, and the estimates are defined as the minimizers of these objective functions. With penalization, some values of $\eta_j^{(k)}$'s can be shrunk to exactly zero, and variables with nonzero $\eta_j^{(k)}$'s are identified as important prognostic markers and associated with the kth cancer type. The magnitudes and signs of $\eta_j^{(k)}$'s describe the strengths and directions of associations. Following the literature, the coordinate descent (CD) technique is adopted for effectively optimizing the objective functions. Details are provided in Appendix A.

The objective function (2) analyzes one gene at a time, and enjoys stable estimation and simple optimization. It may be limited by a lack of attention to the interconnections among genes and their joint effects on cancer prognosis. Our brief literature search suggests that marginal analysis is still highly popular in high-dimensional omics studies [21]. For marginal analysis, a two-stage method is often adopted for marker identification, where multiple tests are first performed and a multiple comparison adjustment is then conducted on p values using, for example, the false discovery rate approach. By contrast with this strategy, we adopt the penalization technique, which can generate more stable results and, more importantly, effectively accommodate the similarity across cancer types. Specifically, MCP is used for regularized estimation and marker identification, which has been shown to have satisfactory theoretical and numerical properties. The most significant advancement is the $\rho\left(\eta_j^{(k)}, \eta_j^{(k')}\right)$ penalty term which promotes similarity between the estimated coefficients of each cancer pair. Data integration is conducted in the discovery process to facilitate early information borrowing. With the magnitude-based shrinkage penalty (3), the magnitudes of gene effects across cancer types are promoted to be similar if they have the same signs, while with the sign-based shrinkage penalty (4), the signs of gene effects are promoted to be similar. Thus, the proposed two types of $\rho\left(\eta_j^{(k)}, \eta_j^{(k')}\right)$ promote different types of similarity, with the former for *quantitative* similarity and the latter for *qualitative* similarity. As in practice the relatedness of cancer types may be not accurately known, both penalties can be useful. λ_1 and λ_2 are two tuning parameters which control the sparsity and similarity of coefficients, respectively. For the p objective functions, we impose the same values of λ_1 and λ_2 on different $\eta_j^{(k)}$ to be concordant with joint analysis. If $\lambda_2 = 0$, the proposed approach goes back to the unintegrated strategy that analyzes each cancer type separately with MCP.

2.2.2. Joint Analysis

For $k = 1, \ldots, K$, consider the AFT model with the joint effects of all omics measurements,

$$T_i^{(k)} = \alpha^{(k)} + X_i^{(k)} \beta^{(k)} + \varepsilon_i^{(k)}, \quad (5)$$

where $\alpha^{(k)}$ is the intercept, $\beta^{(k)} = \left(\beta_1^{(k)}, \ldots, \beta_p^{(k)}\right)'$ is the p-dimensional unknown coefficient vector, and $\varepsilon_i^{(k)}$ is the random error. With the same notations as in the marginal analysis, for estimation, consider the following weighted penalized objective function

$$\sum_{k=1}^{K} \left[\frac{1}{2n^{[k]}} \sum_i w_i^{[k]} \left[y_i^{[k]} - \alpha^{[k]} - X_i^{[k]} \beta^{[k]} \right]^2 \right] + \sum_{k=1}^{K} \sum_{j=1}^{p} \rho_{MCP}\left(\beta_j^{(k)}, \lambda_3, \gamma\right) + \frac{\lambda_4}{2} \sum_{k' \neq k} \sum_{j=1}^{p} \rho\left(\beta_j^{(k)}, \beta_j^{(k')}\right), \quad (6)$$

where λ_3 and λ_4 are the tuning parameters. The KM weights, MCP, and two proposals for $\rho\left(\beta_j^{(k)}, \beta_j^{(k')}\right)$ are also adopted in (6). The proposed estimate is defined as the minimizer of (6). Variables with nonzero estimates are identified as associated with prognosis. For optimization, the CD algorithm is adopted (Appendix A).

Different from (2), objective function (6) jointly analyzes a large number of genes in a single model and thus accommodates a high dimensionality. Compared to marginal analysis, it advances by taking the combined effects of multiple genes into consideration and better describing the underlying disease biology. However, it involves more complex computation and may lead to less stable results. Penalization is adopted to accommodate high dimensionality and identify important genes. It is perhaps the most popular technique in high dimensional data analysis. Different from the existing studies, the magnitude- and sign-based shrinkage penalty terms are also introduced similarly to that in Section 2.2.1. This can effectively accommodate the similarity across cancer types and facilitate information borrowing.

The proposed analysis can be effectively realized. To facilitate data analysis within and beyond this study, we have developed R code and made it publicly available at www.github.com/shuanggema/IntePanCancer.

3. Results

3.1. Marginal Analysis

We analyze the TCGA data using the approach described in Section 2.2.1 with penalties (3) (referred to as A1) and (4) (referred to as A2), as well as an alternative marginal approach A3 which analyzes each cancer type separately with MCP for identifying relevant markers. Comparing with the benchmark A3 can straightforwardly establish the merit of the proposed integrative analysis. Detailed estimation results are provided in the Supplementary Excel file. Different approaches are observed to generate different findings. Specifically, a total of 910 genes with 482 unique ones and 1160 genes with 275 unique ones are identified with A1 and A2, respectively, compared to 2655 genes with 999 unique ones with A3.

In Table 2, we present the top five genes with the largest numbers of associated cancer types and refer to the Supplementary Excel file for more detailed results. It is observed that the numbers of multiple cancer types-related genes identified with A1 and A2 are slightly larger than those with A3. For example, both A1 and A2 identify gene *APH1A* as associated with all nine cancer types, but this gene is missed by A3. Literature search suggests that the identified genes with the proposed A1 and A2 may have important biological implications. For example, the Kyoto Encyclopedia of Genes and Genomes (KEGG) pathway analysis of gene *APH1A* suggests that it is a member of the notch signaling pathway which has an important impact on developmental and cell fate decisions and is deregulated in human solid tumors [22]. APH1A is one of the four essential components of γ-secretase [23]. γ-secretase is a multiprotein intramembrane-cleaving protease, which can cleave ligand-activated endogenous Notch receptors and is a potential drug target for cancer [24]. Gene *MAPK1*, identified as associated with eight cancer types by both A1 and A2, has been reported to be involved in many cancer related pathways. MAPK1 is one of the MAP kinases in the MAPK pathway. It can phosphorylate transcription factors, which regulate the expressions of genes involved in cell proliferation and differentiation. Besides, MAPK1 is involved in EGFR tyrosine kinase inhibitor resistance [25], which importantly contributes to the etiology of various types of cancer, such as pancreatic cancer [26], paediatric acute lymphoblastic leukemia [27], and others. In this process, MAPK1 acts as a serine/threonine kinase upstream of FRS2, which plays a role in epidermal growth factor (EGF) signaling [28]. MAPK1 has also been reported to have an impact on the malignant behavior of breast cancer cells. Published studies show that gene *ETV6*, identified as associated with eight and seven cancer types by A1 and A2, respectively, is involved in the transcriptional dysregulation of cancer pathways. The dysregulation of transcription factors can alter the expressions of target genes and lead to the tumorigenic process. For example, ETV6 is a negative regulator of transcription 3 (Stat3) transcription factor activity, which has the ability to mediate the inhabitation of the proliferation of tumor cells [29]. Gene *ETV6* is relevant to multiple cancer types, including breast cancer [30], leukemia [31], non-small cell lung cancer [32], and others. These biological findings provide support to the validity of the proposed integrative analysis.

Table 2. Marginal analysis: top five genes with the largest numbers of associated cancer types.

Approach	Gene	Number of Associated Cancer Types
A1	APH1A	9
	ETV6	8
	MAPK1	8
	MDS2	8
	AKT2	7
A2	APH1A	9
	CXCR4	8
	MAPK1	8
	ACVR2A	7
	ETV6	7
A3	LAMA1	8
	IGF1	7
	NAPA	7
	TCTA	7
	TNFRSF10D	7

To gain a deeper insight into the identification results, we further calculate the relative overlapping between gene sets associated with different cancer types. Specifically, for two gene sets A and B, their relative overlapping is defined as $\text{ROL}(A, B) = \frac{A \cap B}{A \cup B}$, with a larger value indicating a stronger similarity. Results for different approaches are shown in Table 3. The average ROL values are 0.143 (A1), 0.308 (A2), and 0.147 (A3), respectively, suggesting that A2 leads to gene sets with a higher level of relative overlapping and A1 and A3 have comparable performance. Take breast invasive carcinoma (BRCA) and ovarian serous cystadenocarcinoma (OV), which are established as related, as an example. The ROL values for A1, A2, and A3 are 0.150, 0.265, and 0.146, respectively. The proposed A2 can improve the *qualitative* similarity of genes selected for multiple cancer types to a certain extent.

Table 3. Marginal analysis: relative overlapping between different cancer types.

Approach		BRCA	GBM	HNSC	LAML	LUAD	LUSC	OV	PAAD
A1	BLCA	0.134	0.145	0.135	0.147	0.189	0.070	0.145	0.191
	BRCA		0.167	0.120	0.145	0.148	0.072	0.150	0.119
	GBM			0.149	0.169	0.215	0.089	0.208	0.119
	HNSC				0.202	0.199	0.108	0.154	0.160
	LAML					0.144	0.140	0.134	0.165
	LUAD						0.102	0.141	0.173
	LUSC							0.114	0.058
	OV								0.117

Table 3. Cont.

Approach		BRCA	GBM	HNSC	LAML	LUAD	LUSC	OV	PAAD
A2	BLCA	0.314	0.267	0.320	0.295	0.351	0.161	0.185	0.337
	BRCA		0.272	0.330	0.306	0.234	0.262	0.265	0.341
	GBM			0.416	0.443	0.271	0.310	0.364	0.237
	HNSC				0.346	0.380	0.253	0.366	0.430
	LAML					0.337	0.398	0.302	0.338
	LUAD						0.241	0.298	0.339
	LUSC							0.292	0.201
	OV								0.286
A3	BLCA	0.252	0.082	0.168	0.102	0.196	0.252	0.124	0.162
	BRCA		0.091	0.245	0.101	0.255	0.364	0.146	0.191
	GBM			0.052	0.055	0.065	0.069	0.060	0.071
	HNSC				0.116	0.198	0.251	0.096	0.162
	LAML					0.105	0.108	0.135	0.074
	LUAD						0.227	0.115	0.176
	LUSC							0.134	0.197
	OV								0.081

Beyond identification, we also take a closer look at the estimation results. Specifically, we compute the difference of the estimated coefficient matrices for each cancer pair. Consider the relative Euclidean distance defined as $\sum_{j=1}^{p}\left(\eta_j^{(k)}-\eta_j^{(k')}\right)^2 / \sqrt{\sum_{j=1}^{p}\left(\eta_j^{(k)}\right)^2 \sum_{j=1}^{p}\left(\eta_j^{(k')}\right)^2}$ for $k \neq k'$, with a smaller value indicating a stronger similarity. Results for the three approaches are provided in Table 4, with the average values being 1.606 (A1), 1.534 (A2), and 2.254 (A3). The relative Euclidean distances with A1 and A2 are observed to be smaller than those with A3. For example, the distance values between BRCA and OV are 1.443 with A1 and 1.220 with A2, which are much smaller than 3.230 with A3. As another example, for the two squamous cell carcinomas, lung squamous cell carcinoma (LUSC) and head and neck squamous cell carcinoma (HNSC), the relative Euclidean distances are 1.644 (A1), 1.855 (A2), and 2.577 (A3), respectively. To more intuitively describe similarity, we conduct the hierarchical clustering analysis based on the relative Euclidean distances and present the results in Figure A1 (Appendix B). Biologically sensible findings are made, for example, the distance between BRCA and OV decreases after integration.

Table 4. Marginal analysis: relative Euclidean distances between estimated coefficient matrices.

Approach		BRCA	GBM	HNSC	LAML	LUAD	LUSC	OV	PAAD
A1	BLCA	1.426	1.465	1.572	1.422	1.277	2.160	1.441	1.318
	BRCA		1.270	1.853	1.551	1.457	2.445	1.443	1.584
	GBM			1.974	1.658	1.389	2.722	1.362	1.766
	HNSC				1.205	1.424	1.644	1.530	1.382
	LAML					1.403	1.591	1.471	1.373
	LUAD						1.960	1.463	1.376
	LUSC							1.942	1.959
	OV								1.532
A2	BLCA	1.166	1.250	1.435	1.378	1.075	2.814	1.424	1.070
	BRCA		1.202	1.585	1.482	1.356	2.800	1.220	1.111
	GBM			1.384	1.176	1.293	2.746	1.028	1.307
	HNSC				1.236	1.288	1.855	1.465	1.118
	LAML					1.269	1.764	1.457	1.252
	LUAD						2.347	1.337	1.160
	LUSC							2.512	2.658
	OV								1.205
A3	BLCA	2.354	2.162	1.896	2.029	2.217	2.752	2.203	2.099
	BRCA		2.862	2.364	2.514	1.974	1.870	3.230	1.956
	GBM			2.108	1.929	2.832	2.731	1.835	2.613
	HNSC				1.985	2.151	2.577	2.237	1.916
	LAML					2.455	2.405	2.221	2.207
	LUAD						2.019	3.100	1.988
	LUSC							3.154	2.206
	OV								2.871

3.2. Joint Analysis

Similar to marginal analysis, in joint analysis we adopt both the magnitude-based shrinkage (referred to as B1) and the sign-based shrinkage (referred to as B2). We also consider an alternative joint analysis referred to as B3, which analyzes each cancer type separately and applies MCP to accommodate high dimensionality and select relevant markers. Detailed estimation results are provided in the Supplementary Excel file. For the nine cancer types combined, B1, B2, and B3 identify a total of 1135 genes with 662 unique ones, 1064 genes with 598 unique ones, and 530 genes with 421 unique ones, respectively. The two proposed approaches lead to results different from the alternative. In addition, the joint analysis identification results also differ from those in marginal analysis.

The top five genes with the largest numbers of associated cancer types are provided in Table 5, and more results are provided in the Supplementary Excel file. Similar patterns are observed where the proposed two approaches identify more genes associated with multiple cancer types. For the identified genes, a literature search provides independent evidences of their associations with multiple cancer types. For example, the important biological implications of gene *APH1A* have been already discussed in Section 3.1. In addition, gene *CCAR2*, identified as important for all nine cancer types with B2, has been reported to be associated with the development of many cancer types. It plays a pivotal role in DNA damage response and promoting apoptosis. The depletion of *CCAR2* can impair the activation of the AKT pathway, which ultimately causes the inhibition of cancer cell growth [33]. Specifically, it binds to the BRCA1 C Terminus (BRCT) domain of the tumor suppressor BRCA1 and inhibits BRCA1 in breast cancer [34]. Cho, et al. [35] also suggested that the expression of *CCAR2* is closely related with the progression of ovarian carcinomas. In Kim, et al. [36], an increase in apoptosis was observed in *CCAR2*-deficient non-small cell lung cancer cell lines. Wagle, et al. [37] demonstrated that the expression of *CCAR2* is significantly associated with a higher clinical stage and predicted shorter survival in osteosarcoma. Gene *BTLA* is identified as important for eight cancer types with B2. It is an immunoinhibitory receptor and can deliver inhibitory signals for suppressing lymphocyte activation. The ability of *BTLA* to inhibit tumor-specific human CD8+ T cells suggests it as a target for cancer immunotherapy [38]. Published studies also suggest that gene *BTLA* is relevant to the occurrence and development of many cancer types [39]. For example, a case-control study conducted by Fu, et al. [40] on women from northeast China suggested that breast cancer risk and prognosis may be affected by *BTLA* gene polymorphisms. In addition, Oguro, et al. [41] showed that *BTLA* is closely associated with shorter overall survival in gallbladder cancer. Gene *RUNX2* is identified by B2 as important for five cancer types. The transcription factor RUNX2 can regulate the expressions of genes that are associated with tumor promotion, invasion, and metastasis, such as *VEGF* [42]. *RUNX2* is also involved in many pathways that are related to tumorigenesis, such as the WNT pathway, transforming growth factor beta (TGFβ) signaling pathway, and p53 pathway [42].

Table 5. Joint analysis: top five genes with the largest numbers of associated cancer types.

Approach	Gene	Number of Associated Cancer Types
B1	ETV6	6
	GOT1	6
	CHIC2	5
	CSNK2A1	5
	RUNX2	5
B2	APH1A	9
	CCAR2	9
	HIST1H2AL	9
	BTLA	8
	LAMA1	8

Table 5. Cont.

Approach	Gene	Number of Associated Cancer Types
	EPO	4
	FASLG	4
B3	WDR18	4
	CCND2	3
	CRADD	3

The relative overlapping and Euclidean distances between different cancer types are presented in Tables A1 and A2 (Appendix B). The average values of relative overlapping are 0.103 (B1), 0.107 (B2), and 0.030 (B3), and the average values of Euclidean distance are 2.261 (B1), 1.980 (B2), and 2.459 (B3). Both measures indicate that the proposed joint integrative analysis can improve the identified similarity across cancer types. Take BRCA and PAAD, the relatedness of which has been suggested in literature, as an example. It has been demonstrated that protein annexin A1, A2, A4 and A5 play an important role in the occurrence and development of these two cancer types [43], and *BRCA1* and *BRCA2* gene mutations are commonly observed in both cancer types [44]. The values of relative overlapping are 0.074 (B1), 0.116 (B2), and 0.027 (B3), and the relative Euclidean distances are 1.949 (B1), 1.906 (B2), and 3.829 (B3). For the two common lung cancer subtypes, lung adenocarcinoma (LUAD) and LUSC, the relative overlapping values are 0.098 (B1), 0.119 (B2), and 0.039 (B3), and the relative Euclidean distances are 2.250 (B1), 2.012 (B2), and 2.998 (B3). Results of hierarchical clustering analysis based on the relative Euclidean distances are shown in Figure A2 (Appendix B). With the proposed B1 and B2, cancer types with stronger relatedness tend to be assigned to the same clusters.

Advancing from marginal analysis, joint analysis has the capability of predicting survival time besides marker identification. To evaluate prediction performance, a resampling procedure is adopted. Specifically, for each of the nine cancers, we first split data randomly into a training and a testing set. The training sets for the nine cancer types are then used to fit models and obtain parameter estimates. Finally, we make prediction for the testing set subjects with the estimated parameters. For evaluation, C-statistic is adopted, which is one of the most popular measures for censored survival data [45,46]. It is the integrated AUC (area under the curve) of the time-dependent ROC curve and has value between 0.5 and 1, with a larger value indicating a better prediction performance. The average values over 100 resamplings are shown in Table 6. Overall, B1 and B2 perform better than B3, with B1 having a prominent superiority. For example, for LUSC, the average C-statistic values are 0.748 (B1), 0.649 (B2), and 0.612 (B3). The improvement in prediction accuracy suggests the benefit of integrative analysis of multiple cancer types.

Table 6. Joint analysis: prediction performance of different approaches (mean C-statistic).

	BLCA	BRCA	GBM	HNSC	LAML	LUAD	LUSC	OV	PAAD
B1	0.665	0.876	0.604	0.641	0.573	0.688	0.748	0.577	0.689
B2	0.597	0.719	0.581	0.567	0.551	0.601	0.649	0.562	0.632
B3	0.587	0.693	0.558	0.604	0.558	0.594	0.612	0.547	0.589

3.3. Simulation Based on TCGA Data

To gain more insights into the performance of the proposed integrative analysis, we conduct practical data-based simulation under various scenarios. The specific settings were as follows. (1) The observed gene expression measurements on nine cancer types from TCGA were used as predictors. To generate variations across simulation replicates, we adopted a resampling approach. (2) Set $p = 200$, 500, or 1000. For each value of p, genes were randomly selected from the original gene set. (3) For each cancer type, there were 10 genes associated with the cancer outcomes with nonzero regression coefficients $\beta_{(1)}^{(k)}, \ldots, \beta_{(10)}^{(k)}$. The rest of the coefficients were zeros. (4) For each subject, the event time was computed from the AFT model $\log\left(T_i^{(k)}\right) = \sum_{j=1}^{5} x_{i(j)}^{(k)} \beta_{(j)}^{(k)} + \sum_{j=6}^{10} \left(x_{i(j)}^{(k)}\right)^2 \beta_{(j)}^{(k)} + \varepsilon_i$, where the random error ε_i was generated from $N(0,1)$. Censoring times were randomly generated from an exponential distribution, and the parameter was adjusted to make the censoring rate around 20%. It is noted that to mimic the complexity of real data, the data generating models are more complicated than the simple AFTs with the presence of a small number of quadratic effects. We consider various values of $\beta_{(1)}^{(k)}, \ldots, \beta_{(10)}^{(k)}$ to generate different levels of signal-to-noise ratios and cancer similarity. Under Scenarios I and II, the nine cancer types have the same set of important genes with the same nonzero effects. In particular, for $j = 1, \ldots, 10$ and $k = 1, \ldots, 9$, we set $\beta_{(j)}^{(k)} = 5$ and 2 for Scenarios I and II, respectively. Under Scenario III, the nine cancer types have the same set of important genes, but the magnitudes of effects vary. Specifically, $\beta_{(j)}^{(k)}$'s are randomly generated from $U(1,5)$. Under Scenario IV, the nine cancer types have different sets of important genes. Specifically, the first five important genes have the same effects for all nine cancer types with $\beta_{(j)}^{(k)} = 2$, and the other five important genes are "randomly selected" (and hence likely to differ across datasets) and with $\beta_{(j)}^{(k)} = 2$. There are a total of 12 simulation settings, comprehensively covering different numbers of genes, and different levels of signal-to-noise ratios and cancer similarity.

Analysis was conducted using the proposed marginal and joint analysis approaches as well as two alternatives. To evaluate identification performance, we computed the true positive rate (TPR) and false positive rate (FPR). The average TPR and FPR values over 100 replicates are provided in Table A3, together with the numbers of the identified true positives associated with all nine cancer types (NG). Overall, the four integrative analysis approaches perform better than the two alternatives, with larger values of TPR and smaller values of FPR. For example, under Scenario I with $p = 200$, the average values of (TPR, FPR) are (0.980, 0.258) with A1, (0.951, 0.185) with A2, (0.944, 0.641) with A3, (0.838, 0.087) with B1, (0.880, 0.085) with B2, and (0.688, 0.200) with B3, respectively. The proposed approaches also identify genes with more overlaps across cancer types. Under this specific setting, the average values of NG are 7.0 (A1), 8.4 (A2), 3.8 (A3), 5.7 (B1), 8.8 (B2), and 1.4 (B3). Compared to Scenario I which has a higher signal-to-noise ratio, performance of all six approaches decay under Scenarios II–IV. Similar patterns are observed when dimensionality increases, where all approaches behave worse. However, the proposed approaches still have favorable performance. Take Scenario IV with $p = 500$ as an example, the proposed A1, A2, B1, and B2 have (TPR, FPR) = (0.822, 0.058), (0.678, 0.054), (0.864, 0.040), and (0.719, 0.046), compared to (0.617, 0.116) with A3 and (0.646, 0.038) with B3. In addition, the average values of NG are 4.6 (A1), 2.6 (A2), 0.0 (A3), 5.0 (B1), 3.2 (B2), and 1.8 (B3). As the sign consistency of some genes does not hold under Scenario IV, A2 and B2 have inferior performance compared to A1 and B1, but still have superior performance compared to A3 and B3. The superiority of the proposed integrative analysis approaches observed in data-based simulation provides certain confidence to data analysis results.

4. Discussion

In cancer research, prognosis modeling with omics measurements plays an essential role. The existing studies mostly conduct analysis on one single type of cancer and often suffer from a lack of sufficient information. Integrative analysis represents an emerging trend in recent biomedical studies, among which the most common is the integrative analysis of multiple types of omics data, including gene expressions, copy number variations, and some others, and has led to interesting findings beyond single type omics data-based analysis. In this study, we have taken a different perspective and conducted integrative analysis on multiple cancer types to facilitate across-cancer information borrowing. Similarity across cancer types has been extensively studied in the literature, which provides a solid biological ground for our integrative analysis. Both marginal and joint analysis have been developed with two types of similarity-based penalty, which have intuitive formulations and solid statistical basis. We have analyzed mRNA gene expression data on nine TCGA cancer types with censored survival outcomes. Biologically sensible findings different from the benchmark analysis have been made.

The proposed analysis can be directly applied to other types of omics data and other cancer types. In this study, we have focused on prognosis data and the AFT model. A continuous outcome can be regarded as a special case of prognosis outcome without censoring, and thus the proposed analysis can be applied directly. It can also be extended to accommodate categorical outcomes using, for example, generalized linear models. With the availability of multiple types of omics data on multiple cancer types, it can be of interest to conduct the two types of integration simultaneously. More functional examination of the data analysis results will be needed to confirm the findings.

Supplementary Materials: The following are available online at http://www.mdpi.com/2073-4425/10/8/604/s1: Detailed results referred to in Section 3 are available in the Supplementary Excel file. Table S1: Detailed estimation and identification results.

Author Contributions: All authors contributed to conceptualization, methodology, and writing. S.W. conducted data analysis.

Funding: This research was partly funded by the National Institutes of Health [CA216017, CA204120]; National Natural Science Foundation of China [91546202, 71331006]; Bureau of Statistics of China [2018LD02]; "Chenguang Program" supported by Shanghai Education Development Foundation and Shanghai Municipal Education Commission [18CG42]; and Program for Innovative Research Team of Shanghai University of Finance and Economics.

Acknowledgments: We are very grateful to the reviewers for their careful review and insightful comments, which have led to a significant improvement of this article.

Conflicts of Interest: The authors declare no conflict of interest. The funders had no role in the design of the study; in the collection, analyses, or interpretation of data; in the writing of the manuscript; or in the decision to publish the results.

Appendix A

For optimizing objective functions (2) and (6), a weighted normalization is first conducted as:

$$y_i^{(k)} = \sqrt{w_i^{(k)}}\left(y_i^{(k)} - \overline{y}^{(k)}\right), \quad x_{ij}^{(k)} = \sqrt{w_i^{(k)}}\left(x_{ij}^{(k)} - \overline{x}_j^{(k)}\right),$$

where $\overline{y}^{(k)} = \sum_{i=1}^{n^{(k)}} w_i^{(k)} y_i^{(k)} / \sum_{i=1}^{n^{(k)}} w_i^{(k)}$ and $\overline{x}_j^{(k)} = \sum_{i=1}^{n^{(k)}} w_i^{(k)} x_{ij}^{(k)} / \sum_{i=1}^{n^{(k)}} w_i^{(k)}$. Then objective functions (2) and (6) can be rewritten as:

$$\sum_{k=1}^{K}\left[\frac{1}{2n^{[k]}}\sum_i \left[y_i^{[k]} - x_{ij}^{[k]}\eta_j^{[k]}\right]^2\right] + \sum_{k=1}^{K}\rho_{MCP}\left(\eta_j^{(k)},\lambda_1,\gamma\right) + \frac{\lambda_2}{2}\sum_{k'\neq k}^{K}\rho\left(\eta_j^{(k)},\eta_j^{(k')}\right), \quad (A1)$$

and

$$\sum_{k=1}^{K}\left[\frac{1}{2n^{[k]}}\sum_i \left[y_i^{[k]} - X_i^{[k]}\beta^{[k]}\right]^2\right] + \sum_{k=1}^{K}\sum_{j=1}^{p}\rho_{MCP}\left(\beta_j^{(k)},\lambda_3,\gamma\right) + \frac{\lambda_4}{2}\sum_{k'\neq k}\sum_{j=1}^{p}\rho\left(\beta_j^{(k)},\beta_j^{(k')}\right). \quad (A2)$$

The coordinate descent (CD) technique is used to optimize objective functions (A1) and (A2). In the CD procedure, the objective function is optimized with respect to one parameter at a time, and the other parameters are fixed at their current values. All parameters are iteratively cycled through until convergence.

Specifically, with fixed tuning parameters, for $j = 1,\ldots,p$, the CD algorithm for penalized objective function (A1) proceeds as follows.

(1). Initialize $t = 0$, $\left(\eta_j^{(k)}\right)^{(t)} = 0$, $k = 1,..,K$, where $\left(\eta_j^{(k)}\right)^{(t)}$ denotes the estimate of $\eta_j^{(k)}$ at iteration t.

(2). For $k = 1,\ldots,K$, carry out the following steps sequentially.

(2.1) If $\rho\left(\eta_j^{(k)}, \eta_j^{(k')}\right)$ is the magnitude-based shrinkage penalty (3), compute:

$$b = \frac{1}{n^{(k)}} \sum_{i=1}^{n^{(k)}} x_{ij}^{(k)2} + \lambda_2 \text{ and } a = \frac{1}{n^{(k)}} \sum_{i=1}^{n^{(k)}} x_{ij}^{(k)} y_i^{(k)} + \lambda_2 \sum_{k' \neq k} s_j^{(kk')} \left(\eta_j^{(k')}\right)^{(t)}.$$

If $\rho\left(\eta_j^{(k)}, \eta_j^{(k')}\right)$ is the sign-based shrinkage penalty (4), compute:

$$b = \frac{1}{n^{(k)}} \sum_{i=1}^{n^{(k)}} x_{ij}^{(k)2} + \frac{\lambda_2}{\left(\left(\eta_j^{(k)}\right)^{(t)} + \chi\right)^2},$$

$$a = \frac{1}{n^{(k)}} \sum_{i=1}^{n^{(k)}} x_{ij}^{(k)} y_i^{(k)} + \lambda_2 \sum_{k' \neq k} \frac{\left(\eta_j^{(k')}\right)^{(t)}}{\left(\left(\eta_j^{(k)}\right)^{(t)} + \chi\right)\left(\left(\eta_j^{(k')}\right)^{(t)} + \chi\right)},$$

where χ is a small positive number, which is set as 0.01 in our numerical study.

(2.2) If $\left|\frac{a}{b}\right| > \gamma \lambda_1$, update $\left(\eta_j^{(k)}\right)^{(t+1)} = \frac{a}{b}$;

else if $|a| > \lambda_1$, update $\left(\eta_j^{(k)}\right)^{(t+1)} = \frac{a - \text{Sgn}(a) * \lambda_1}{(b-1)/\gamma}$;

else, update $\left(\eta_j^{(k)}\right)^{(t+1)} = 0$.

(3). Repeat Step (2) until convergence. In our numerical study, convergence is concluded if $\sum_{k=1}^{K} \left| \left\| \eta_j^{|k|} \right\|^{|t+1|} - \left\| \eta_j^{|k|} \right\|^{|t|} \right| < 10^{-4}$.

With fixed tuning parameters, the CD algorithm for penalized objective function (A2) proceeds as follows.

(1). Initialize $t = 0$, $\left(\beta^{(k)}\right)^{(t)} = (0, \ldots, 0)'$, $k = 1, \ldots, K$, where $\left(\beta^{(k)}\right)^{(t)}$ denotes the estimate of $\beta^{(k)}$ at iteration t.

(2). For $j = 1, \ldots, p$ and $k = 1, \ldots, K$, carry out the following steps sequentially.

(2.1) If $\rho\left(\beta_j^{(k)}, \beta_j^{(k')}\right)$ is the magnitude-based shrinkage penalty (3), compute:

$$b = \frac{1}{n^{(k)}} \sum_{i=1}^{n^{(k)}} x_{ij}^{(k)2} + \lambda_4, \text{ and } a = \frac{1}{n^{(k)}} \sum_{i=1}^{n^{(k)}} x_{ij}^{(k)} \left(y_i^{(k)} - \sum_{j' \neq j}^{p} x_{ij'}^{(k)} \beta_{j'}^{(k)}\right) +$$
$$\lambda_2 \sum_{k' \neq k} s_j^{(kk')} \left(\beta_j^{(k')}\right)^{(t)}.$$

If $\rho\left(\beta_j^{(k)}, \beta_j^{(k')}\right)$ is the sign-based shrinkage penalty (4), compute:

$$b = \frac{1}{n^{(k)}} \sum_{i=1}^{n^{(k)}} x_{ij}^{(k)2} + \frac{\lambda_4}{\left(\left(\beta_j^{(k)}\right)^{(t)} + \chi\right)^2},$$

$$a = \frac{1}{n^{(k)}} \sum_{i=1}^{n^{(k)}} x_{ij}^{(k)} \left(y_i^{(k)} - \sum_{j' \neq j}^{p} x_{ij'}^{(k)} \beta_{j'}^{(k)}\right) + \lambda_2 \sum_{k' \neq k} \frac{\left(\beta_j^{(k')}\right)^{(t)}}{\left(\left(\beta_j^{(k)}\right)^{(t)} + \chi\right)\left(\left(\beta_j^{(k')}\right)^{(t)} + \chi\right)},$$

where χ is a small positive number, which is set as 0.01 in our numerical study.

(2.2) If $\left|\frac{a}{b}\right| > \gamma\lambda_3$, update $\left(\beta_j^{(k)}\right)^{(t+1)} = \frac{a}{b}$;

else if $|a| > \lambda_3$, update $\left(\beta_j^{(k)}\right)^{(t+1)} = \frac{a - Sgn(a) * \lambda_3}{(b-1)/\gamma}$;

else, update $\left(\beta_j^{(k)}\right)^{(t+1)} = 0$.

(3). Repeat Step (2) until convergence. In our numerical study, convergence is concluded if $\sum_{j=1}^{p} \sum_{k=1}^{K} \left|\left(\beta_j^{(k)}\right)^{(t+1)} - \left(\beta_j^{(k)}\right)^{(t)}\right| < 10^{-4}$.

These approaches involve tuning parameters, which are selected using cross validation.

Appendix B

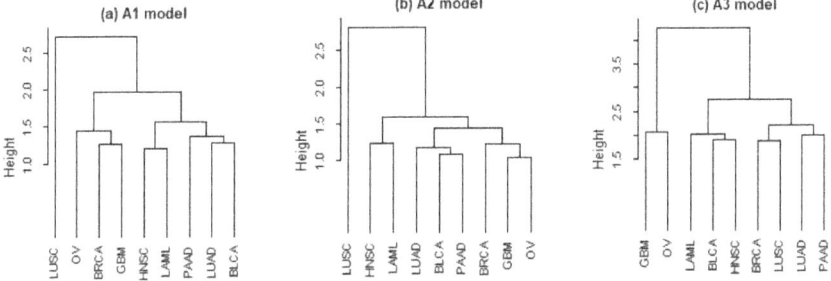

Figure A1. Marginal analysis: clustering dendrogram based on the relative Euclidean distances.

Table A1. Joint analysis: relative overlapping between different cancer types.

Approach		BRCA	GBM	HNSC	LAML	LUAD	LUSC	OV	PAAD
B1	BLCA	0.075	0.087	0.114	0.124	0.116	0.126	0.087	0.105
	BRCA		0.068	0.069	0.152	0.086	0.082	0.099	0.074
	GBM			0.138	0.129	0.085	0.106	0.089	0.114
	HNSC				0.114	0.121	0.086	0.112	0.071
	LAML					0.137	0.11	0.129	0.088
	LUAD						0.098	0.114	0.083
	LUSC							0.127	0.097
	OV								0.091
B2	BLCA	0.097	0.085	0.090	0.128	0.124	0.121	0.124	0.107
	BRCA		0.095	0.088	0.104	0.091	0.102	0.133	0.116
	GBM			0.101	0.113	0.071	0.109	0.124	0.127
	HNSC				0.130	0.090	0.101	0.124	0.109
	LAML					0.098	0.102	0.138	0.09
	LUAD						0.119	0.097	0.089
	LUSC							0.115	0.114
	OV								0.132
B3	BLCA	0.034	0.026	0.01	0.024	0.02	0.046	0.039	0.038
	BRCA		0.012	0.014	0.011	0.026	0.016	0.041	0.027
	GBM			0.015	0.033	0.008	0.042	0.068	0.000
	HNSC				0.000	0.000	0.038	0.018	0.017
	LAML					0.084	0.015	0.03	0.063
	LUAD						0.039	0.052	0.028
	LUSC							0.057	0.018
	OV								0.073

Table A2. Joint analysis: relative Euclidean distances between estimated coefficient matrices.

Approach		BRCA	GBM	HNSC	LAML	LUAD	LUSC	OV	PAAD
B1	BLCA	2.108	2.090	2.503	1.943	1.994	2.104	2.122	2.081
	BRCA		2.262	2.164	2.063	2.04	2.787	2.474	1.949
	GBM			2.454	2.082	2.002	2.266	2.001	2.230
	HNSC				2.331	2.538	3.571	2.846	1.960
	LAML					2.047	2.383	2.114	2.079
	LUAD						2.250	1.958	2.147
	LUSC							2.093	2.878
	OV								2.481
B2	BLCA	1.983	1.931	1.973	1.794	1.807	2.122	1.908	1.771
	BRCA		1.928	1.891	1.963	2.063	2.423	2.093	1.906
	GBM			1.838	1.890	1.921	1.986	1.967	1.875
	HNSC				1.965	2.098	2.371	2.095	1.832
	LAML					1.843	1.889	1.866	1.940
	LUAD						2.012	1.953	1.880
	LUSC							2.064	2.351
	OV								2.071
B3	BLCA	3.664	2.176	1.992	2.251	2.052	3.432	2.185	2.049
	BRCA		2.672	3.223	2.528	3.074	1.994	2.672	3.829
	GBM			2.049	2.029	2.029	2.759	2.017	2.219
	HNSC				2.124	2.004	3.088	2.099	2.040
	LAML					1.994	2.455	1.978	1.907
	LUAD						2.998	1.983	2.101
	LUSC							2.720	3.722
	OV								2.421

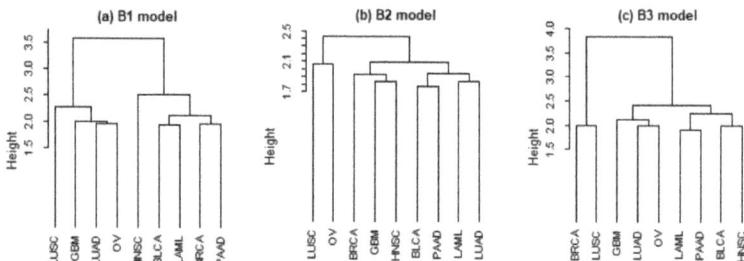

Figure A2. Joint analysis: clustering dendrogram based on the relative Euclidean distances.

Table A3. Data-based simulation: average true positive rates (TPRs) and false positive rates (FPRs) of different approaches, and numbers of identified true positives associated with all nine cancer types (NG).

p	Scenario		A1	A2	A3	B1	B2	B3
200	I	TPR	0.980	0.951	0.944	0.838	0.880	0.688
		FPR	0.258	0.185	0.641	0.087	0.085	0.200
		NG	7.0	8.4	3.8	5.7	8.8	1.4
	II	TPR	0.697	0.681	0.678	0.735	0.691	0.533
		FPR	0.263	0.172	0.537	0.231	0.169	0.347
		NG	4.4	3.7	0.4	4.6	4.0	0.0
	III	TPR	0.841	0.801	0.752	0.821	0.813	0.565
		FPR	0.258	0.297	0.303	0.312	0.321	0.422
		NG	7.0	6.0	5.7	5.6	6.3	1.4
	IV	TPR	0.843	0.741	0.621	0.897	0.766	0.662
		FPR	0.124	0.176	0.195	0.072	0.053	0.052
		NG	3.3	2.3	0.0	5.0	3.0	2.1

Table A3. Cont.

p	Scenario		A1	A2	A3	B1	B2	B3
500	I	TPR	0.922	0.911	0.844	0.933	0.844	0.688
		FPR	0.248	0.152	0.452	0.114	0.122	0.173
		NG	5.0	8.0	3.0	5.0	5.0	0.0
	II	TPR	0.672	0.664	0.653	0.647	0.643	0.647
		FPR	0.191	0.171	0.165	0.025	0.063	0.128
		NG	4.7	2.8	0.4	3.8	3.1	0.0
	III	TPR	0.774	0.723	0.445	0.811	0.784	0.644
		FPR	0.173	0.160	0.107	0.173	0.053	0.181
		NG	4.0	6.3	0.0	6.2	4.8	1.2
	IV	TPR	0.822	0.678	0.617	0.864	0.719	0.646
		FPR	0.058	0.054	0.116	0.042	0.046	0.038
		NG	4.6	2.6	0.0	5.0	3.2	1.8
1000	I	TPR	0.733	0.722	0.623	0.622	0.688	0.591
		FPR	0.198	0.173	0.350	0.001	0.056	0.064
		NG	5.0	6.0	3.0	3.0	3.0	0.0
	II	TPR	0.674	0.643	0.622	0.689	0.689	0.611
		FPR	0.161	0.075	0.136	0.011	0.108	0.061
		NG	2.1	4.0	0.0	2.0	3.0	0.0
	III	TPR	0.664	0.667	0.624	0.692	0.677	0.564
		FPR	0.038	0.069	0.297	0.096	0.043	0.076
		NG	4.0	6.4	0.6	5.2	5.4	0.4
	IV	TPR	0.722	0.644	0.622	0.855	0.711	0.699
		FPR	0.093	0.100	0.136	0.016	0.015	0.009
		NG	5.0	5.0	0.0	5.0	3.0	2.0

References

1. Nicholson, R.I.; Gee, J.M.; Harper, M.E. EGFR and cancer prognosis. *Eur. J. Cancer* **2001**, *37* (Suppl. 4), S9–S15. [CrossRef]
2. Petitjean, A.; Achatz, M.I.; Borresen-Dale, A.L.; Hainaut, P.; Olivier, M. TP53 mutations in human cancers: functional selection and impact on cancer prognosis and outcomes. *Oncogene* **2007**, *26*, 2157–2165. [CrossRef] [PubMed]
3. Gao, J.; Aksoy, B.A.; Dogrusoz, U.; Dresdner, G.; Gross, B.; Sumer, S.O.; Sun, Y.; Jacobsen, A.; Sinha, R.; Larsson, E.; et al. Integrative analysis of complex cancer genomics and clinical profiles using the cBioPortal. *Sci. Signal.* **2013**, *6*. [CrossRef] [PubMed]
4. Chiu, C.G.; Nakamura, Y.; Chong, K.K.; Huang, S.K.; Kawas, N.P.; Triche, T.; Elashoff, D.; Kiyohara, E.; Irie, R.F.; Morton, D.L.; et al. Genome-wide characterization of circulating tumor cells identifies novel prognostic genomic alterations in systemic melanoma metastasis. *Clin. Chem.* **2014**, *60*, 873–885. [CrossRef] [PubMed]
5. Hoadley, K.A.; Yau, C.; Hinoue, T.; Wolf, D.M.; Lazar, A.J.; Drill, E.; Shen, R.; Taylor, A.M.; Cherniack, A.D.; Thorsson, V.; et al. Cell-of-Origin patterns dominate the molecular classification of 10,000 tumors from 33 types of cancer. *Cell* **2018**, *173*, 291–304.e6. [CrossRef] [PubMed]
6. Vogelstein, B.; Kinzler, K.W. Cancer genes and the pathways they control. *Nat. Med.* **2004**, *10*, 789–799. [CrossRef] [PubMed]
7. Easton, D.F.; Ford, D.; Bishop, D.T. Breast and ovarian cancer incidence in BRCA1-mutation carriers. Breast Cancer Linkage Consortium. *Am. J. Hum. Genet.* **1995**, *56*, 265–271.
8. Jin, G.; Kim, M.J.; Jeon, H.S.; Choi, J.E.; Kim, D.S.; Lee, E.B.; Cha, S.I.; Yoon, G.S.; Kim, C.H.; Jung, T.H.; et al. PTEN mutations and relationship to EGFR, ERBB2, KRAS, and TP53 mutations in non-small cell lung cancers. *Lung Cancer* **2010**, *69*, 279–283. [CrossRef]

9. Hammerman, P.S.; Sos, M.L.; Ramos, A.H.; Xu, C.; Dutt, A.; Zhou, W.; Brace, L.E.; Woods, B.A.; Lin, W.; Zhang, J.; et al. Mutations in the DDR2 kinase gene identify a novel therapeutic target in squamous cell lung cancer. *Cancer Discov.* **2011**, *1*, 78–89. [CrossRef]
10. Dutt, A.; Ramos, A.H.; Hammerman, P.S.; Mermel, C.; Cho, J.; Sharifnia, T.; Chande, A.; Tanaka, K.E.; Stransky, N.; Greulich, H.; et al. Inhibitor-sensitive FGFR1 amplification in human non-small cell lung cancer. *PLoS ONE* **2011**, *6*, e20351. [CrossRef]
11. Cava, C.; Bertoli, G.; Colaprico, A.; Olsen, C.; Bontempi, G.; Castiglioni, I. Integration of multiple networks and pathways identifies cancer driver genes in pan-cancer analysis. *BMC Genom.* **2018**, *19*, 25. [CrossRef]
12. Yu, X.; Lian, B.; Wang, L.; Zhang, Y.; Dai, E.; Meng, F.; Liu, D.; Wang, S.; Liu, X.; Wang, J.; et al. The pan-cancer analysis of gene expression patterns in the context of inflammation. *Mol. Biosyst.* **2014**, *10*, 2270–2276. [CrossRef]
13. Sharma, A.; Jiang, C.; De, S. Dissecting the sources of gene expression variation in a pan-cancer analysis identifies novel regulatory mutations. *Nucleic Acids Res.* **2018**, *46*, 4370–4381. [CrossRef]
14. Martinez-Ledesma, E.; Verhaak, R.G.; Trevino, V. Identification of a multi-cancer gene expression biomarker for cancer clinical outcomes using a network-based algorithm. *Sci. Rep.* **2015**, *5*, 11966. [CrossRef]
15. Leiserson, M.D.; Vandin, F.; Wu, H.T.; Dobson, J.R.; Eldridge, J.V.; Thomas, J.L.; Papoutsaki, A.; Kim, Y.; Niu, B.; McLellan, M.; et al. Pan-cancer network analysis identifies combinations of rare somatic mutations across pathways and protein complexes. *Nat. Genet.* **2015**, *47*, 106–114. [CrossRef]
16. Xing, L.; Lesperance, M.; Zhang, X. Simultaneous prediction of multiple outcomes using revised stacking algorithms. *Bioinformatics* **2019**. [CrossRef]
17. Matlock, K.; De Niz, C.; Rahman, R.; Ghosh, S.; Pal, R. Investigation of model stacking for drug sensitivity prediction. *BMC Bioinform.* **2018**, *19*, 71. [CrossRef]
18. Zhang, D.; Shen, D. Multi-modal multi-task learning for joint prediction of multiple regression and classification variables in Alzheimer's disease. *NeuroImage* **2012**, *59*, 895–907. [CrossRef]
19. TruSight RNA Pan-Cancer Panel. Available online: https://support.illumina.com/sequencing/sequencing_kits/trusight-rna-pan-cancer-panel/questions.html (accessed on 7 March 2019).
20. Huang, J.; Ma, S.; Xie, H. Regularized estimation in the accelerated failure time model with high-dimensional covariates. *Biometrics* **2006**, *62*, 813–820. [CrossRef]
21. Zhang, Y.; Dai, Y.; Zheng, T.; Ma, S. Risk Factors of Non-Hodgkin Lymphoma. *Expert Opin. Med. Diagn.* **2011**, *5*, 539–550. [CrossRef]
22. Takebe, N.; Nguyen, D.; Yang, S.X. Targeting notch signaling pathway in cancer: clinical development advances and challenges. *Pharmacol. Ther.* **2014**, *141*, 140–149. [CrossRef] [PubMed]
23. Zhao, G.; Liu, Z.; Ilagan, M.X.; Kopan, R. Gamma-secretase composed of PS1/Pen2/Aph1a can cleave notch and amyloid precursor protein in the absence of nicastrin. *J. Neurosci.* **2010**, *30*, 1648–1656. [CrossRef] [PubMed]
24. Miele, L.; Miao, H.; Nickoloff, B.J. NOTCH signaling as a novel cancer therapeutic target. *Curr. Cancer Drug Targets* **2006**, *6*, 313–323. [CrossRef] [PubMed]
25. Laag, E.; Majidi, M.; Cekanova, M.; Masi, T.; Takahashi, T.; Schuller, H.M. NNK activates ERK1/2 and CREB/ATF-1 via beta-1-AR and EGFR signaling in human lung adenocarcinoma and small airway epithelial cells. *Int. J. Cancer* **2006**, *119*, 1547–1552. [CrossRef] [PubMed]
26. Furukawa, T.; Kanai, N.; Shiwaku, H.O.; Soga, N.; Uehara, A.; Horii, A. AURKA is one of the downstream targets of MAPK1/ERK2 in pancreatic cancer. *Oncogene* **2006**, *25*, 4831–4839. [CrossRef]
27. Yan, J.; Jiang, N.; Huang, G.; Tay, J.L.; Lin, B.; Bi, C.; Koh, G.S.; Li, Z.; Tan, J.; Chung, T.H.; et al. Deregulated MIR335 that targets MAPK1 is implicated in poor outcome of paediatric acute lymphoblastic leukaemia. *Br. J. Haematol.* **2013**, *163*, 93–103. [CrossRef]
28. Wu, Y.; Chen, Z.; Ullrich, A. EGFR and FGFR signaling through FRS2 is subject to negative feedback control by ERK1/2. *Biol. Chem.* **2003**, *384*, 1215–1226. [CrossRef]
29. Milde-Langosch, K.; Bamberger, A.M.; Rieck, G.; Grund, D.; Hemminger, G.; Muller, V.; Loning, T. Expression and prognostic relevance of activated extracellular-regulated kinases (ERK1/2) in breast cancer. *Br. J. Cancer* **2005**, *92*, 2206–2215. [CrossRef]
30. Li, Z.; Tognon, C.E.; Godinho, F.J.; Yasaitis, L.; Hock, H.; Herschkowitz, J.I.; Lannon, C.L.; Cho, E.; Kim, S.J.; Bronson, R.T.; et al. ETV6-NTRK3 fusion oncogene initiates breast cancer from committed mammary progenitors via activation of AP1 complex. *Cancer Cell* **2007**, *12*, 542–558. [CrossRef]

31. Bohlander, S.K. ETV6: A versatile player in leukemogenesis. *Semin. Cancer Biol.* **2005**, *15*, 162–174. [CrossRef]
32. Liang, J.Z.; Li, Y.H.; Zhang, Y.; Wu, Q.N.; Wu, Q.L. Expression of ETV6/TEL is associated with prognosis in non-small cell lung cancer. *Int. J. Clin. Exp. Pathol.* **2015**, *8*, 2937–2945.
33. Restelli, M.; Magni, M.; Ruscica, V.; Pinciroli, P.; De Cecco, L.; Buscemi, G.; Delia, D.; Zannini, L. A novel crosstalk between CCAR2 and AKT pathway in the regulation of cancer cell proliferation. *Cell Death Dis.* **2016**, *7*, e2453. [CrossRef]
34. Hiraike, H.; Wada-Hiraike, O.; Nakagawa, S.; Koyama, S.; Miyamoto, Y.; Sone, K.; Tanikawa, M.; Tsuruga, T.; Nagasaka, K.; Matsumoto, Y.; et al. Identification of DBC1 as a transcriptional repressor for BRCA1. *Br. J. Cancer* **2010**, *102*, 1061–1067. [CrossRef]
35. Cho, D.; Park, H.; Park, S.H.; Kim, K.; Chung, M.; Moon, W.; Kang, M.; Jang, K. The expression of DBC1/CCAR2 is associated with poor prognosis of ovarian carcinoma. *J. Ovarian Res.* **2015**, *8*, 2. [CrossRef]
36. Kim, W.; Jeong, J.W.; Kim, J.E. CCAR2 deficiency augments genotoxic stress-induced apoptosis in the presence of melatonin in non-small cell lung cancer cells. *Tumour Biol.* **2014**, *35*, 10919–10929. [CrossRef]
37. Wagle, S.; Park, S.H.; Kim, K.M.; Moon, Y.J.; Bae, J.S.; Kwon, K.S.; Park, H.S.; Lee, H.; Moon, W.S.; Kim, J.R.; et al. DBC1/CCAR2 is involved in the stabilization of androgen receptor and the progression of osteosarcoma. *Sci. Rep.* **2015**, *5*, 13144. [CrossRef]
38. Derre, L.; Rivals, J.P.; Jandus, C.; Pastor, S.; Rimoldi, D.; Romero, P.; Michielin, O.; Olive, D.; Speiser, D.E. BTLA mediates inhibition of human tumor-specific CD8+ T cells that can be partially reversed by vaccination. *J. Clin. Investig.* **2010**, *120*, 157–167. [CrossRef]
39. Haymaker, C.; Wu, R.; Bernatchez, C.; Radvanyi, L. PD-1 and BTLA and CD8(+) T-cell "exhaustion" in cancer: "Exercising" an alternative viewpoint. *Oncoimmunology* **2012**, *1*, 735–738. [CrossRef]
40. Fu, Z.; Li, D.; Jiang, W.; Wang, L.; Zhang, J.; Xu, F.; Pang, D.; Li, D. Association of BTLA gene polymorphisms with the risk of malignant breast cancer in Chinese women of Heilongjiang Province. *Breast Cancer Res. Treat.* **2010**, *120*, 195–202. [CrossRef]
41. Oguro, S.; Ino, Y.; Shimada, K.; Hatanaka, Y.; Matsuno, Y.; Esaki, M.; Nara, S.; Kishi, Y.; Kosuge, T.; Hiraoka, N. Clinical significance of tumor-infiltrating immune cells focusing on BTLA and Cbl-b in patients with gallbladder cancer. *Cancer Sci.* **2015**, *106*, 1750–1760. [CrossRef]
42. Cohen-Solal, K.A.; Boregowda, R.K.; Lasfar, A. RUNX2 and the PI3K/AKT axis reciprocal activation as a driving force for tumor progression. *Mol. Cancer* **2015**, *14*, 137. [CrossRef]
43. Deng, S.; Wang, J.; Hou, L.; Li, J.; Chen, G.; Jing, B.; Zhang, X.; Yang, Z. Annexin A1, A2, A4 and A5 play important roles in breast cancer, pancreatic cancer and laryngeal carcinoma, alone and/or synergistically. *Oncol. Lett.* **2013**, *5*, 107–112. [CrossRef]
44. Stadler, Z.K.; Salo-Mullen, E.; Patil, S.M.; Pietanza, M.C.; Vijai, J.; Saloustros, E.; Hansen, N.A.; Kauff, N.D.; Kurtz, R.C.; Kelsen, D.P.; et al. Prevalence of BRCA1 and BRCA2 mutations in Ashkenazi Jewish families with breast and pancreatic cancer. *Cancer* **2012**, *118*, 493–499. [CrossRef]
45. Schroder, M.S.; Culhane, A.C.; Quackenbush, J.; Haibe-Kains, B. survcomp: An R/Bioconductor package for performance assessment and comparison of survival models. *Bioinformatics* **2011**, *27*, 3206–3208. [CrossRef]
46. Jiang, Y.; Shi, X.; Zhao, Q.; Krauthammer, M.; Rothberg, B.E.; Ma, S. Integrated analysis of multidimensional omics data on cutaneous melanoma prognosis. *Genomics* **2016**, *107*, 223–230. [CrossRef]

© 2019 by the authors. Licensee MDPI, Basel, Switzerland. This article is an open access article distributed under the terms and conditions of the Creative Commons Attribution (CC BY) license (http://creativecommons.org/licenses/by/4.0/).

Article

Model-Based Clustering with Measurement or Estimation Errors

Wanli Zhang † and Yanming Di *

Department of Statistics, Oregon State University, Corvallis, OR 97330, USA; zhang_wan_li@lilly.com
* Correspondence: diy@stat.oregonstate.edu
† Current address: Eli Lilly & Company, Shanghai 200021, China.

Received: 27 November 2019; Accepted: 5 February 2020; Published: 10 February 2020

Abstract: Model-based clustering with finite mixture models has become a widely used clustering method. One of the recent implementations is MCLUST. When objects to be clustered are summary statistics, such as regression coefficient estimates, they are naturally associated with estimation errors, whose covariance matrices can often be calculated exactly or approximated using asymptotic theory. This article proposes an extension to Gaussian finite mixture modeling—called MCLUST-ME—that properly accounts for the estimation errors. More specifically, we assume that the distribution of each observation consists of an underlying true component distribution and an independent measurement error distribution. Under this assumption, each unique value of estimation error covariance corresponds to its own classification boundary, which consequently results in a different grouping from MCLUST. Through simulation and application to an RNA-Seq data set, we discovered that under certain circumstances, explicitly, modeling estimation errors, improves clustering performance or provides new insights into the data, compared with when errors are simply ignored, whereas the degree of improvement depends on factors such as the distribution of error covariance matrices.

Keywords: gaussian finite mixture model; clustering analysis; uncertainty; expectation-maximization algorithm; classification boundary; gene expression; RNA-seq

1. Introduction

Model-based clustering [1,2] is one of the most commonly used clustering methods. The authors of [3] introduced the methodology of clustering objects through analyzing a mixture of distributions. The main assumption is that objects within a class share a common distribution in their characteristics, whereas objects from a different class will follow a different distribution. The entire population will then follow a mixture of distributions, and the purpose of clustering would be to take such a mixture and analyze it into simple components and estimate the "probabilities of membership", that is, the probabilities that each observation belongs to each cluster.

One of the most recent implementations of model-based clustering is MCLUST [4–6], in which each observation is assumed to follow a finite mixture of multivariate Gaussian distributions. MCLUST describes cluster geometries (shape, volume, and orientation) by reparameterizing component covariance matrices [7], and formulates different models by imposing constraints on each geometric feature. The expectation-maximization (EM) algorithm [8,9] is used for maximum likelihood estimation, and the Bayesian information criterion (BIC) [10,11] is used for selection of optimal model(s).

In most cases, observations to be clustered are assumed to have been precisely measured, whereas there are situations where this assumption is clearly not feasible. This article proposes an extension to Gaussian mixture modeling that properly accounts for measurement or estimation errors in the special case when the error distributions are either known or can be estimated, as well as introduces the clustering algorithm built upon it, which we named MCLUST-ME. The real data example that motivated our study is where we apply clustering algorithm to coefficients from gene-wise regression analysis of an RNA-seq data set (see Section 3.3 for details). For each gene, five of the fitted regression coefficients correspond to log fold changes in mean expression levels between two groups of *Arabidopsis* plants at five time points after treatment. In such a case, we can reasonably approximate the error covariances of the regression coefficients by inverting the observed information matrix. In general, whenever one applies clustering analysis to a set of summary statistics, it is often possible to approximate the distribution of their estimation errors with, for instance, Gaussian distributions. In this paper, we describe how the estimation/measurement errors, with known or estimated error covariances, can be incorporated into the model-based clustering framework. An obvious alternative strategy in practice is to ignore the individual estimation/measurement errors. We will use simulations and the real data example to understand in what circumstances explicitly modeling the estimation errors will improve the clustering results, and to what degree.

In Section 3.3, we will compare the results of applying the MCLUST method and our new MCLUST-ME method to cluster the log fold changes of 1000 randomly selected genes from the RNA-seq data set mentioned above at two of the time points where the gene expressions were most active. Here, we briefly summarize the input data structure for the clustering analysis and the highlights of the results. Columns 2 and 3 of Table 1 list the estimated log fold changes and their standard errors for 15 representative genes at the two time points being analyzed: these are from the regression analysis applied to each row (gene) of the RNA-seq data set. (The standard errors are the square roots of the corresponding diagonal entries of the error covariances.) In particular, we included 10 genes that are classified differently by the MCLUST and the MCLUST-ME methods. We note that there is sizable variation among the standard errors of the log fold changes. When MCLUST was used to cluster the log fold changes, the estimation errors will be ignored: As long as two genes have the same log fold changes at the two time points, they will always belong to the same cluster. However, we understand that, in this context, a moderate log fold change with a high estimation error is less significant than the same log fold change with low estimation error: this would be obvious if we were to perform a hypothesis test for differential expression (DE), but existing clustering methods such as MCLUST cannot readily incorporate such information into a clustering analysis. The MCLUST-ME method we propose in this paper aims to incorporate information about the estimation errors into the clustering analysis. One distinctive feature of the new MCLUST-ME method is that two points with similar log fold changes may not belong to the same cluster: it also depends on the error covariances of the log fold changes. We note that when the 1000 genes were classified into two clusters by MCLUST and MCLUST-ME, the two clusters for this data set roughly correspond to a "DE" cluster and a "non-DE" cluster. Columns 4 and 5 of Table 1 list the probabilities to the "non-DE" cluster estimated by MCLUST and by MCLUST-ME. We see that the genes that were classified into the "non-DE" cluster by MCLUST-ME, but to the "DE" cluster by MCLUST tend to be genes having moderate log fold changes, but relatively large error covariances. We do not have the ground truth for this data set, but the results from the new MCLUST-ME method alert us that not all log fold changes are created equal.

The organization of the rest of this article is as follows. Section 2 briefly reviews the MCLUST method and then introduces our extension, MCLUST-ME. In particular, Section 2.6 investigates decision boundaries of the two methods for two-group clustering. Sections 3.1 and 3.2 give simulation settings and results on comparing MCLUST-ME with MCLUST in terms classification accuracy and uncertainty. Section 3.3 gives an example where we cluster a real-life data set using both methods. Finally, conclusions and perspectives for future work are addressed in Section 4.

Table 1. Estimated log fold changes at two time points, associated standard errors, and estimated membership probabilities to the "non-DE" cluster by MCLUST and by MCLUST-ME, for 15 genes selected from the real-data example. Column 2 and 3 are estimated log fold changes at 1 h and 3 h and their standard errors. Column 4 and 5 are estimated membership probabilities to the "non-DE" cluster by MCLUST and by MCLUST-ME. The first 5 rows are randomly selected from 1000 genes that we analyzed. The second 5 rows are selected among the genes that are classified to the "non-DE" cluster by MCLUST, but to the "DE" cluster by MCLUST-ME: the standard errors of the log fold changes tend to be low in this group; the last 5 rows are selected among genes that are classified to the "DE" cluster by MCLUST, but to the "non-DE" cluster by MCLUST-ME: the standard errors of the log fold changes tend to be high in this group.

Gene ID	Log Fold Change (SE) 1h	Log Fold Change (SE) 3h	z_1 MCLUST	z_1 MCLUST–ME
AT2G42230	−0.277 (0.006)	0.152 (0.006)	0.920	0.921
AT3G56110	0.081 (0.121)	0.228 (0.099)	0.919	0.895
AT1G23330	0.351 (0.018)	−0.209 (0.012)	0.862	0.870
AT5G23060	−0.243 (0.005)	−0.909 (0.005)	0.684	0.751
AT5G06240	−0.680 (0.022)	0.103 (0.012)	0.774	0.764
AT3G20350	−0.952 (0.007)	−0.090 (0.009)	0.562	0.396
AT1G30440	−1.056 (0.010)	−0.398 (0.009)	0.511	0.375
AT1G30490	−0.983 (0.008)	−0.322 (0.006)	0.612	0.480
AT1G23400	−1.017 (0.011)	−0.275 (0.006)	0.547	0.418
AT1G17980	0.734 (0.001)	−0.001 (0.006)	0.524	0.363
AT2G30890	−1.040 (0.150)	−0.142 (0.125)	0.445	0.771
AT5G15160	−0.044 (0.129)	−1.059 (0.225)	0.332	0.866
AT5G45310	−0.221 (0.162)	−1.404 (0.313)	0.042	0.837
AT5G46871	0.373 (0.065)	0.886 (0.094)	0.305	0.581
AT2G22240	0.076 (0.016)	−0.975 (0.043)	0.371	0.690

2. Materials and Methods

2.1. Review of MCLUST Model

Finite mixture model Let $f_1(\boldsymbol{y};\boldsymbol{\Theta}_1), f_2(\boldsymbol{y};\boldsymbol{\Theta}_2), ..., f_G(\boldsymbol{y};\boldsymbol{\Theta}_G)$ be G probability distributions defined on the d-dimensional random vector $\boldsymbol{y}$, and a mixture of the G distributions is formed by taking proportions $\{\tau_k\}$ of the population from components $\{f_k\}$, with probability density given by

$$f(\boldsymbol{y};\boldsymbol{\Theta}) = \sum_{k=1}^{G} \tau_k f_k(\boldsymbol{y};\boldsymbol{\Theta}_k), \tag{1}$$

where $\boldsymbol{\Theta} = (\boldsymbol{\Theta}_1, ..., \boldsymbol{\Theta}_G)$ are model parameters.

Component density The MCLUST model assumes that the distribution of each $\boldsymbol{y}$ is a mixture of multivariate normal distributions. Under the MCLUST model, the component density of $\boldsymbol{y}$ in group k is

$$f_k(\boldsymbol{y};\boldsymbol{\mu}_k,\boldsymbol{\Sigma}_k) = \frac{\exp\left\{-\frac{1}{2}(\boldsymbol{y}-\boldsymbol{\mu}_k)^T\boldsymbol{\Sigma}_k^{-1}(\boldsymbol{y}-\boldsymbol{\mu}_k)\right\}}{\sqrt{\det[2\pi\boldsymbol{\Sigma}_k]}}, \tag{2}$$

In other words,

$$\boldsymbol{y}|k \sim N_d(\boldsymbol{\mu}_k,\boldsymbol{\Sigma}_k). \tag{3}$$

The (marginal) probability density of $\boldsymbol{y}$ is given by

$$f(\boldsymbol{y}) = \sum_{k=1}^{G} \tau_k f_k(\boldsymbol{y};\boldsymbol{\mu}_k,\boldsymbol{\Sigma}_k). \tag{4}$$

Likelihood function   Suppose a sample of n independent and identically distributed (iid) random vectors $\boldsymbol{y} = (\boldsymbol{y}_1, ..., \boldsymbol{y}_n)$ is drawn from the mixture. The (observed) log likelihood of the sample is then

$$l_O(\boldsymbol{\Theta}; \boldsymbol{y}) = \sum_{i=1}^{n} \log f(\boldsymbol{y}_i) = \sum_{i=1}^{n} \log \sum_{k=1}^{G} \tau_k f_k(\boldsymbol{y}_i; \boldsymbol{\mu}_k, \boldsymbol{\Sigma}_k), \tag{5}$$

where $\boldsymbol{\Theta} = (\tau_1, ..., \tau_G; \boldsymbol{\mu}_1, ..., \boldsymbol{\mu}_G; \boldsymbol{\Sigma}_1, ..., \boldsymbol{\Sigma}_G)$ are the model parameters.

2.2. MCLUST-ME Model

We extend the MCLUST model by associating each data point with an error term and assumes that the covariance matrix of each error term is either known or can be estimated.

Component density   Given that $\boldsymbol{y}$ belongs to component k, the MCLUST-ME models assumes that there exists a latent variable $\boldsymbol{w}$, representing its "truth" part, and $\boldsymbol{\epsilon}$, representing its "error" part, such that

$$\begin{cases} \boldsymbol{y} = \boldsymbol{w} + \boldsymbol{\epsilon}, \\ \boldsymbol{w}|k \sim N_d(\boldsymbol{\mu}_k, \boldsymbol{\Sigma}_k), \\ \boldsymbol{\epsilon} \sim N_d(\boldsymbol{0}, \boldsymbol{\Lambda}), \end{cases} \tag{6}$$

where $\boldsymbol{w}$ and $\boldsymbol{\epsilon}$ are independent. $\boldsymbol{\mu}_k$ and $\boldsymbol{\Sigma}_k$ are unknown mean and covariance parameters (same as in the MCLUST model), and $\boldsymbol{\Lambda}$ is the known error covariance matrix associated with $\boldsymbol{y}$. The distribution of $\boldsymbol{y}$ being in component k is then

$$\boldsymbol{y}|k \sim N_d(\boldsymbol{\mu}_k, \boldsymbol{\Sigma}_k + \boldsymbol{\Lambda}), \tag{7}$$

with density function

$$g_k(\boldsymbol{y}; \boldsymbol{\mu}_k, \boldsymbol{\Sigma}_k, \boldsymbol{\Lambda}) = \frac{\exp\left\{-\frac{1}{2}(\boldsymbol{y} - \boldsymbol{\mu}_k)^T (\boldsymbol{\Sigma}_k + \boldsymbol{\Lambda})^{-1} (\boldsymbol{y} - \boldsymbol{\mu}_k)\right\}}{\sqrt{\det[2\pi(\boldsymbol{\Sigma}_k + \boldsymbol{\Lambda})]}}, \tag{8}$$

and the (marginal) probability density of $\boldsymbol{y}$ is given by

$$g(\boldsymbol{y}) = \sum_{k=1}^{G} \tau_k g_k(\boldsymbol{y}; \boldsymbol{\mu}_k, \boldsymbol{\Sigma}_k, \boldsymbol{\Lambda}). \tag{9}$$

Likelihood function   Suppose a sample of n iid random vectors $\boldsymbol{y} = (\boldsymbol{y}_1, ..., \boldsymbol{y}_n)$ is drawn from the mixture, where each $\boldsymbol{y}_i$ is associated with known error covariance matrix $\boldsymbol{\Lambda}_i$. The (observed) log likelihood of the sample is then

$$l_O(\boldsymbol{\Theta}; \boldsymbol{y}) = \sum_{i=1}^{n} \log g(\boldsymbol{y}_i) = \sum_{i=1}^{n} \log \sum_{k=1}^{G} \tau_k g_k(\boldsymbol{y}_i; \boldsymbol{\mu}_k, \boldsymbol{\Sigma}_k, \boldsymbol{\Lambda}_i), \tag{10}$$

where $\boldsymbol{\Theta} = (\tau_1, ..., \tau_G; \boldsymbol{\mu}_1, ..., \boldsymbol{\mu}_G; \boldsymbol{\Sigma}_1, ..., \boldsymbol{\Sigma}_G)$ are the model parameters.

In summary, the MCLUST-ME and MCLUST models have the same set of model parameters for the normal components and the mixing proportions. The key difference is that under the MCLUST-ME model, the measurement or observation errors of the observations are explicitly modeled, and observations are each associated with a given error covariance matrix.

2.3. Expectation-Maximization (EM) Algorithm

In the original MCLUST method, the EM algorithm is used to estimate the unknown parameters and compute the membership probabilities. In this subsection, we will first review the EM algorithm under the general MCLUST framework, and then highlight the differences in implementation between the MCLUST method and the MCLUST-ME method.

Complete data log likelihood Given observations $(\mathbf{y}_1, ..., \mathbf{y}_n)$, suppose that each $\mathbf{y}_i$ is associated with one of G states. Then, there exists unobserved indicator vectors $\{\mathbf{z}_i = (z_{i1}, ..., z_{iG})\}$ where $\mathbf{z}_i \overset{iid}{\sim} \text{Mult}_G(1, \boldsymbol{\tau})$ with $\boldsymbol{\tau} = (\tau_1, ..., \tau_G)$. The complete data then consists of $\mathbf{x}_i = (\mathbf{y}_i, \mathbf{z}_i)$. Assuming that the conditional probability density of $\mathbf{y}_i$ given $\mathbf{z}_i$ is $\prod_{k=1}^{G} f_k(\mathbf{y}_i; \boldsymbol{\mu}_k, \boldsymbol{\Sigma}_k, \boldsymbol{\Lambda}_i)^{z_{ik}}$, the complete data log likelihood can be derived as follows,

$$\begin{aligned}
l_C &= \log \prod_{i=1}^{n} f(\mathbf{y}_i, \mathbf{z}_i) \\
&= \log \prod_{i=1}^{n} f(\mathbf{y}_i | \mathbf{z}_i) f(\mathbf{z}_i) \\
&= \log \prod_{i=1}^{n} \left[\prod_{k=1}^{G} f_k(\mathbf{y}_i; \boldsymbol{\mu}_k, \boldsymbol{\Sigma}_k, \boldsymbol{\Lambda}_i)^{z_{ik}} \right] \left[\prod_{k=1}^{G} \tau_k^{z_{ik}} \right] \\
&= \log \prod_{i=1}^{n} \prod_{k=1}^{G} [\tau_k f_k(\mathbf{y}_i; \boldsymbol{\mu}_k, \boldsymbol{\Sigma}_k, \boldsymbol{\Lambda}_i)]^{z_{ik}} \\
&= \sum_{i=1}^{n} \sum_{k=1}^{G} z_{ik} \log [\tau_k f_k(\mathbf{y}_i; \boldsymbol{\mu}_k, \boldsymbol{\Sigma}_k, \boldsymbol{\Lambda}_i)].
\end{aligned} \qquad (11)$$

EM iterations The EM algorithm consists of iterations of an *M step* and an *E step*, as described below.

- *M step*: Given current estimates of $\{z_{ik}\}$, maximize the complete-data log-likelihood l_C with respect to $(\tau_k, \boldsymbol{\mu}_k, \boldsymbol{\Sigma}_k)$.
- *E step*: Given estimates $(\hat{\tau}_k, \hat{\boldsymbol{\mu}}_k, \hat{\boldsymbol{\Sigma}}_k)$ from last *M step*, for all $i = 1, ..., n$ and $k = 1, ..., G$, compute the membership probabilities

$$\hat{z}_{ik} = \frac{\hat{\tau}_k f_k(\mathbf{y}_i; \hat{\boldsymbol{\mu}}_k, \hat{\boldsymbol{\Sigma}}_k, \boldsymbol{\Lambda}_i)}{\sum_{j=1}^{G} \hat{\tau}_j f_j(\mathbf{y}_i; \hat{\boldsymbol{\mu}}_j, \hat{\boldsymbol{\Sigma}}_j, \boldsymbol{\Lambda}_i)}. \qquad (12)$$

The two steps alternate until the increment in l_O is small enough. Upon convergence, a membership probability matrix is produced and each observation is assigned to the most probable cluster, that is,

$$\text{membership of } \mathbf{y}_i = \operatorname{argmax}_k \{\hat{z}_{ik}\}, \qquad (13)$$

and the classification uncertainty for $\mathbf{y}_i$ is defined as

$$1 - \max_k \{\hat{z}_{ik}\}. \qquad (14)$$

In two-group clustering, the classification uncertainty cannot exceed 0.5 (otherwise the point is incorrectly assigned).

For MCLUST, the component density f_k is defined in (2), and for MCLUST-ME, f_k is substituted by g_k in (8).

M-step implementation details For likelihood maximization in the *M step*, a closed-form solution always exists for $\hat{\tau}_k, k = 1, \ldots, G$ (see [12] for more details):

$$\hat{\tau}_k = \frac{1}{n} \sum_{i=1}^{n} z_{ik} \qquad (15)$$

We can derive the estimation equations for $\boldsymbol{\mu}_k$ and $\boldsymbol{\Sigma}_k$ ($k = 1, \ldots, G$) by taking the partial derivatives of l_C with respect to $\boldsymbol{\mu}_k$ and $\boldsymbol{\Sigma}_k$ and setting the derivatives to $\mathbf{0}$. For MCLUST-ME (see [13] for a summary of useful matrix calculus formulas, in particular (11.7) and (11.8)):

$$\frac{\partial l_C}{\partial \boldsymbol{\mu}_k} = \sum_{i=1}^{n} z_{ik}(\boldsymbol{\Sigma}_k + \boldsymbol{\Lambda}_i)^{-1}(\boldsymbol{y}_i - \boldsymbol{\mu}_k) = \mathbf{0} \qquad (16)$$

and

$$\frac{\partial l_C}{\partial \boldsymbol{\Sigma}_k} = \frac{1}{2}\sum_{i=1}^{n} z_{ik}(\boldsymbol{\Sigma}_k + \boldsymbol{\Lambda}_i)^{-1}(\boldsymbol{y}_i - \boldsymbol{\mu}_k)(\boldsymbol{y}_i - \boldsymbol{\mu}_k)^T(\boldsymbol{\Sigma}_k + \boldsymbol{\Lambda}_i)^{-1} - \frac{1}{2}\sum_{i=1}^{n} z_{ik}(\boldsymbol{\Sigma}_k + \boldsymbol{\Lambda}_i)^{-1} = \mathbf{0}. \qquad (17)$$

For estimation equations under the MCLUST model, one set all the $\boldsymbol{\Lambda}_i$'s to $\mathbf{0}$ in the above two equations. Note that under MCLUST, if there is no constraint on $\boldsymbol{\Sigma}_k$, there are closed-form solutions for $\boldsymbol{\mu}_k$ and $\boldsymbol{\Sigma}_k$:

$$\hat{\boldsymbol{\mu}}_k = \frac{\sum_{i=1}^{n} z_{ik}\boldsymbol{y}_i}{\sum_{i=1}^{n} z_{ik}} \qquad (18)$$

and

$$\hat{\boldsymbol{\Sigma}}_k = \frac{\sum_{i=1}^{n} z_{ik}(\boldsymbol{y}_i - \hat{\boldsymbol{\mu}}_k)(\boldsymbol{y}_i - \hat{\boldsymbol{\mu}}_k)^T}{\sum_{i=1}^{n} z_{ik}}. \qquad (19)$$

For MCLUST-ME, each $\boldsymbol{y}_i$ corresponds to a different $\boldsymbol{\Lambda}_i$. One can see that, in general, there is no closed-form solution for $(\boldsymbol{\mu}_k, \boldsymbol{\Sigma}_k)$. In our implementation of the MCLUST-ME M step, we solve $\boldsymbol{\mu}_k$ from (16),

$$\hat{\boldsymbol{\mu}}_k = \left[\sum_{i=1}^{n} z_{ik}(\boldsymbol{\Sigma}_k + \boldsymbol{\Lambda}_i)^{-1}\right]^{-1} \sum_{i=1}^{n} z_{ik}(\boldsymbol{\Sigma}_k + \boldsymbol{\Lambda}_i)^{-1}\boldsymbol{y}_i, \qquad (20)$$

and plug it into (17), and then use the limited-memory BFGS, a quasi-Newton method in R (function optim [14]), to obtain an optimal solution for $\boldsymbol{\Sigma}_k$ numerically. We obtain $\hat{\boldsymbol{\mu}}_k$ by substituting the resulting $\hat{\boldsymbol{\Sigma}}_k$ into (20).

The complexity of the EM algorithm for the MCLUST-ME increase with the number of clusters, the number of parameters (which is determined by the dimension of the data), and the number of observations. It is much slower than the original MCLUST algorithm due to the fact we have to use a numerical optimization routine to find the maximum likelihood estimate (MLE) of $\boldsymbol{\mu}_k$'s and $\boldsymbol{\Sigma}_k$'s in the M step. (See Conclusion and Discussion for a brief summary of running time of MCLUST-ME on the real data example.)

2.4. Initial Values

Owing to its iterative nature, the EM algorithm can start with either an *E step* or an *M step*. In the context of model-based clustering, initiation with the *M step* takes advantage of the availability of other existing clustering methods, in the sense that, given a data set, we can acquire their initial memberships by first clustering the data with other methods. MCLUST adopts model-based agglomerative hierarchical clustering [7,15] to generate initial memberships. Model-based hierarchical clustering aims at maximizing the *classification likelihood* instead of (5) or (10); at each stage, the maximum-likelihood pair of clusters are merged together. Although the resulting partitions are suboptimal due to its heuristic nature, model-based hierarchical clustering has been shown to often yield reasonable results and is relatively easy to compute [16]. In light of this, we also use model-based hierarchical clustering to obtain initial memberships for MCLUST-ME. For the choice of initial values when starting with *E step* (i.e., initial parameter estimates), see [17] for a nice discussion.

2.5. Model Selection

Within MCLUST framework, selection for the number of clusters can be achieved through the use of the Bayesian information criterion (BIC). Given a random sample of n independent d-vectors $\boldsymbol{y} = (\boldsymbol{y}_1, ..., \boldsymbol{y}_n)$ drawn from (4) and (9) with some value of G, the BIC for this G-component mixture model is given by:

$$BIC_G = 2l_O(\hat{\boldsymbol{\Theta}}; \boldsymbol{y}) - \nu_G \log(n), \tag{21}$$

where $\hat{\boldsymbol{\Theta}}$ is the MLE for model parameters, l_O is the observed likelihood as in (5) or (10), and ν_G is the number of independent parameters to be estimated. In the most simplistic case, we allow the mean and covariance of each component to vary freely—this is the case we will focus on in this paper. Therefore, for a G-component mixture model, we have $\nu_G = (G-1) + Gd + Gd(d-1)/2$. For comparison purpose, in this paper, we will compare MCLUST-ME results to MCLUST results with the same number of components.

2.6. Decision Boundaries for Two-Group Clustering

In this subsection, we examine decision boundaries produced by MCLUST and MCLUST-ME for partitioning a sample into $G = 2$ clusters.

2.6.1. MCLUST Boundary

Suppose we would like to separate a d-dimensional i.i.d. random sample $S = \{\boldsymbol{y}_i\}_{i=1}^N$ into two clusters with MCLUST. Let $(\hat{\tau}_k, \hat{\boldsymbol{\mu}}_k, \hat{\boldsymbol{\Sigma}}_k)$ denote MLEs for $(\tau_k, \boldsymbol{\mu}_k, \boldsymbol{\Sigma}_k)$ upon convergence. If we assign each point to the more probable cluster, then the two clusters can be expressed as follows.

$$E_1 = \{\boldsymbol{y}_i \in S : \hat{\tau}_1 f_1(\boldsymbol{y}_i; \hat{\boldsymbol{\mu}}_1, \hat{\boldsymbol{\Sigma}}_1) - \hat{\tau}_2 f_2(\boldsymbol{y}_i; \hat{\boldsymbol{\mu}}_2, \hat{\boldsymbol{\Sigma}}_2) > 0\}; \quad E_2 = S \setminus E_1, \tag{22}$$

and the decision boundary separating E_1 and E_2 is

$$B = \{\boldsymbol{t} \in \mathbb{R}^d : \hat{\tau}_1 f_1(\boldsymbol{t}; \hat{\boldsymbol{\mu}}_1, \hat{\boldsymbol{\Sigma}}_1) - \hat{\tau}_2 f_2(\boldsymbol{t}; \hat{\boldsymbol{\mu}}_2, \hat{\boldsymbol{\Sigma}}_2) = 0\}, \tag{23}$$

where f_k, $k = 1, 2$, is defined in (2). Equivalently, the boundary B is the set of all points in $\mathbb{R}^d$ with classification uncertainty equal to 0.5. Notice that since the solution set B does not depend on i, a common boundary is shared by *all* observations. When $d = 2$, under the model assumption of MCLUST, the boundary B is a straight line when $\hat{\boldsymbol{\Sigma}}_1 = \hat{\boldsymbol{\Sigma}}_2$, and a conic section when $\hat{\boldsymbol{\Sigma}}_1 \neq \hat{\boldsymbol{\Sigma}}_2$, with its shape and position determined by the values of the MLEs. This can be shown by simplifying the equality in (23) (see [12] for more details).

2.6.2. MCLUST-ME Boundary

Consider the data $S = \{\boldsymbol{y}_i\}_{i=1}^N$ and each $\boldsymbol{y}_i$ is associated with known error covariance $\boldsymbol{\Lambda}_i$ for all i. Suppose our goal is to partition S into two clusters. Let $(\tilde{\tau}_k, \tilde{\boldsymbol{\mu}}_k, \tilde{\boldsymbol{\Sigma}}_k)$ be MLEs from the MCLUST-ME model. If we assign each observation to the more probable cluster, the two clusters can be expressed as follows,

$$E_1^* = \{\boldsymbol{y}_i \in S : \tilde{\tau}_1 g_1(\boldsymbol{y}_i; \tilde{\boldsymbol{\mu}}_1, \tilde{\boldsymbol{\Sigma}}_1, \boldsymbol{\Lambda}_i) - \tilde{\tau}_2 g_2(\boldsymbol{y}_i; \tilde{\boldsymbol{\mu}}_2, \tilde{\boldsymbol{\Sigma}}_2, \boldsymbol{\Lambda}_i) > 0\}; \quad E_2^* = S \setminus E_1^*,$$

where g_k is defined in (8). The above decision rule (and therefore boundary) of classifying each point $\boldsymbol{y}_i$ now depends not only on the values of MLEs, but also on the error covariance matrix, $\boldsymbol{\Lambda}_i$, of $\boldsymbol{y}_i$. Instead of producing a common boundary for all points in S, the MCLUST-ME model specifies an individualized classification boundary for each $\boldsymbol{y}_i$ as follows,

$$B^*(\boldsymbol{\Lambda}_i) = \{\boldsymbol{t} \in \mathbb{R}^d : \tilde{\tau}_1 g_1(\boldsymbol{t}; \tilde{\boldsymbol{\mu}}_1, \tilde{\boldsymbol{\Sigma}}_1, \boldsymbol{\Lambda}_i) - \tilde{\tau}_2 g_2(\boldsymbol{t}; \tilde{\boldsymbol{\mu}}_2, \tilde{\boldsymbol{\Sigma}}_2, \boldsymbol{\Lambda}_i) = 0\}.$$

Similar to our argument in Section 2.6.1, when $d = 2$, $B^*(\Lambda_i)$ is either a straight line or a conic section.

When $\Lambda_i = \Lambda_j$ for some $i \neq j$, that is, when two points are associated with the same error covariance, it can be seen that $B^*(\Lambda_i) = B^*(\Lambda_j)$, meaning that the two points share a common classification boundary. In the special case where $\Lambda_i = \Lambda_j\, \forall i \neq j$, all boundaries $B^*(\Lambda_i)$ will coincide with each other.

One consequence of the existence of multiple decision boundaries is that the classification uncertainty of each point will depend on its corresponding value of Λ_i. In MCLUST, points with high uncertainty (≈ 0.5) are aligned around the single classification boundary, whereas in MCLUST-ME, each highly uncertain point is close to its own boundary. Consequently, as we will see in Section 3.1, our method allows intermixing of points belonging to different clusters, while MCLUST creates clear-cut separation between clusters.

2.7. Related Methods

The authors of [18] discussed a clustering method for data with measurement errors. They also assumed that each observation, y_i, is associated with a known covariance matrix, $\tilde{\Lambda}_i$, but they assume that this covariance matrix is for the distance *between the observation and the center of a cluster*. Their conceptual model, using our notation, assumes that

$$y_i | k \sim N_d(\mu_k, \tilde{\Lambda}_i) \qquad (24)$$

when observation i belongs to cluster k (under their model, group membership is deterministic, not probabilistic). Comparing (24) to our MCLUST-ME model (6) and (7), we see that their model lacks the "model-based" element—the covariance matrix Σ_k—for each cluster k, $k = 1, \ldots, G$. In other words, their $\tilde{\Lambda}_i$ plays the role of our $\Sigma_k + \Lambda_i$. This is a crucial difference: we understand that in MCLUST and MCLUST-ME models, Σ_k's are used to capture different shapes, orientations, and scales of the different clusters. Also, although it is reasonable to assume that the error covariances of the measurements (Λ_i in MCLUST-ME) are known or can be estimated, it is much more difficult to know $\Sigma_k + \Lambda_i$ (i.e., $\tilde{\Lambda}_i$), as we do not where the centers of the clusters are before running the clustering algorithm.

The authors of that paper discussed two heuristic algorithms for fitting G clusters into observations: hError and kError. Under their model, they need to estimate the μ_k's for all the clusters and the deterministic (or hard) group memberships for each observation. Both algorithms are distance-based, and not based on an EM algorithm. The hError algorithm is a hierarchical clustering algorithm: it iteratively merges two current clusters with the smallest distances. The error covariances $\tilde{\Lambda}_i$ were incorporated into the distance formula. For each current cluster k, let S_k be the collection of observations. The center of cluster k is estimated by a weighted average of the observations:

$$\hat{\mu}_k = \Big(\sum_{i \in S_k} \tilde{\Lambda}_i^{-1}\Big)^{-1} \sum_{i \in S_k} \tilde{\Lambda}_i^{-1} y_i \qquad (25)$$

with covariance matrix

$$\Psi_k = \mathrm{Var}(\hat{\mu}_k) = \Big(\sum_{i \in S_k} \tilde{\Lambda}_i^{-1}\Big)^{-1}. \qquad (26)$$

The distance between any two clusters k and l is defined by

$$d_{kl} = (\hat{\mu}_k - \hat{\mu}_l)^T (\Psi_k + \Psi_l)^{-1} (\hat{\mu}_k - \hat{\mu}_l) \qquad (27)$$

The kError algorithm is an extension of the k-means method. It iterates between two steps: (1) Computing the centers of the clusters using (25). (2) Assigning each point to the closest cluster based on the distance formula

$$d_{ik} = (y_i - \hat{\mu}_k)^T \tilde{\Lambda}_i^{-1} (y_i - \hat{\mu}_k). \qquad (28)$$

We implemented the simpler kError algorithm as described above and applied it the real-data example. We summarized our findings in Section 3.3.

The authors of [19] proposed another extension to the *k*-means method that incorporates errors on individual observations. Under their model, each cluster is characterized by a "profile" $\boldsymbol{\alpha} = (\alpha_1, \ldots, \alpha_m)$, where m is the dimension of the data. Each observation, $\boldsymbol{g}_i = (g_{i1}, \ldots, g_{im})$, from this cluster is modeled as

$$g_{ij} = \beta_i \alpha_j + \gamma_i + \epsilon_{ij}, \quad j = 1, \ldots, m, \tag{29}$$

where $\epsilon_{ij} \sim N(0, \sigma_{ij})$ with known error variances σ_{ij}. The distance from an observation $\boldsymbol{g}_i$ to a cluster with profile $\boldsymbol{\alpha}$ is defined as

$$\min_{\beta_i, \gamma_i} \sum_{j=1}^{m} \left[\frac{g_{ij} - (\beta_i \alpha_j + \gamma_i)}{\sigma_{ij}} \right]^2, \tag{30}$$

essentially the weighted sum of squared errors from a weighted least-squares regression of $\boldsymbol{g}_i$ on the profile $\boldsymbol{\alpha}$. The motivation of this distance measure is that it captures both the euclidean distance and the correlation between an observation and a profile. Their version of *k*-means algorithm, CORE, proceeds by iteratively estimating the profile $\boldsymbol{\alpha}$ for each cluster and then assigning each observation $\boldsymbol{g}_i$ to the closest cluster according to (30). We note that their distance measure is less useful for low-dimensional data, as a regression line needs to be fitted between each observation and the cluster profile. If we force the slope β_i to be 0, then we see that their method will be similar to the kError method in [18].

3. Results

In our simulations and real-data example, version 5.0.1 of MCLUST was used.

3.1. Simulation 1: Clustering Performance

We simulated data from bivariate normal mixture distribution with different parameter settings, and applied both MCLUST-ME and MCLUST to partition the data into two clusters. The purpose of this simulation is twofold: first, to investigate the degree of improvement in clustering performance by incorporating known error distributions, and second, to study how error structure affects clustering result.

3.1.1. Data Generation

The data were generated from a two-component bivariate normal mixture distribution, where each point is either error-free or associated with some known, constant error covariance. The data generation process is as follows.

(1) Generate $\{h_i\}_{i=1}^{n}$ i.i.d. from Bernoulli(η). For each i, h_i will serve as indicator for error, and on average, a proportion η of data points will be associated with error.
(2) Generate $\{z_i\}_{i=1}^{n}$ i.i.d. from Bernoulli(τ). Parameter τ will be the mixing proportion.
(3) For $i = 1, \ldots, n$, generate $\boldsymbol{y}_i$ from

$$z_i N_2(\boldsymbol{\mu}_1, \boldsymbol{\Sigma}_1 + h_i \boldsymbol{\Lambda}) + (1 - z_i) N_2(\boldsymbol{\mu}_2, \boldsymbol{\Sigma}_2 + h_i \boldsymbol{\Lambda}).$$

Values of the above parameters are as follows: $\boldsymbol{\mu}_1 = (0,0)^T$, $\boldsymbol{\mu}_2 = (8,0)^T$, $\boldsymbol{\Sigma}_1 = 64 I_2$, $\boldsymbol{\Sigma}_2 = 16 I_2$, $n = 300$, $\tau_1 = \tau_2 = 0.5$, and $\boldsymbol{\Lambda} = 36 I_2$. As the values of z_i provide us with the true memberships of each observation, we are able to use them to evaluate externally the performance of clustering methods in consideration.

3.1.2. Simulation Procedure

The simulation proceeds as follows.

(1) Choose a value for η from $\{0.1, 0.3, 0.5, 0.7, 0.9\}$.
(2) Randomly select a random seed.
(3) Generate a random sample following Section 3.1.1.
(4) Run MCLUST and MCLUST-ME, fixing $G = 2$. Initiate with true memberships.
(5) Repeat (2)–(4) for 100 different seeds.
(6) Repeat (1)–(5) for each value of η.

The membership for each observation as well as MLEs upon convergence will be recorded.

3.1.3. The Adjusted Rand Index

In this simulation study, as the true memberships of the observations are available, we can externally evaluate the performance of both clustering methods by calculating the Rand index [20]. Given n observations and two partitions R and Q of the data, we can use a contingency table (Table 2) to demonstrate their agreement.

Table 2. 2×2 contingency table for comparing partitions R and Q.

Partition	Q	
R	Pair in same group	Pair in different groups
Pair in Same Group	a	b
Pair in Different Groups	c	d

The Rand index (RI) is defined as

$$RI = \frac{a+d}{a+b+c+d}.$$

There are some pitfalls of the Rand index: for two random partitions, the expected value of RI is not equal to zero, and the value of RI tends to one as the number of partitions increases [21]. To overcome these problems, Hubert and Arabie [22] proposed the adjusted Rand index (ARI), which has an expectation of zero. The ARI is defined as

$$ARI = \frac{RI - \text{Expected}(RI)}{1 - \text{Expected}(RI)} = \frac{\binom{n}{2}(a+d) - [(a+b)(a+c) + (c+d)(b+d)]}{\binom{n}{2}^2 - [(a+b)(a+c) + (c+d)(b+d)]}.$$

ARI takes values between -1 and 1, with an ARI of 1 indicating perfect agreement between two partitions (i.e., $RI = 1$), and an ARI of 0 indicating independence between partitions (i.e., $RI = \text{Expected}(RI)$).

Permutation tests can be used to test whether the observed ARI is significantly greater than zero [23]. Although keeping the numbers of partitions and partition sizes the same as the original data, a large number of pairs of partitions are generated at random and ARI is computed for each generated pair. A randomization p-value can then be calculated based on the distribution of generated ARI's. Similarly, permutation p-values can be obtained for testing whether paired ARI values originating from two clustering methods are equal or not.

3.1.4. Simulation 1 Results

Decision boundary We first visualize the clustering results from both methods, as well as the theoretical decision boundaries stated in Section 2.6. Figure 1 shows groupings of the same data generated with $\eta = 0.5$ and with random seed 7.

For MCLUST-ME, we identify two distinct decision boundaries: The dotted curve separates points measured *with* errors (solid) into two groups, whereas the dashed curve separates points *without* errors (empty). For MCLUST, one boundary separates all points, regardless of their associated errors. This confirms our findings in Section 2.6.

For this particular simulation, we make two interesting discoveries. First, the two MCLUST-ME boundaries are relatively far apart. Second, none of the three boundaries intersect with each other. As mentioned in Section 2.6.1, the shape and position of these boundaries completely depend upon corresponding values of MLEs, which, in turn, are end results of a procedure of iterative nature (the EM algorithm). We have additional plots similar to Figure 1 for other values of η and other random seeds in [12].

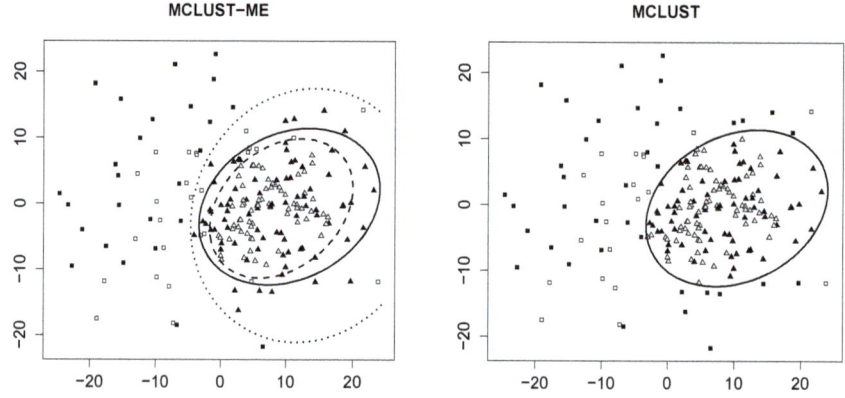

Figure 1. Clustering result of the sample generated with random seed = 7 and $\eta = 0.5$. *Both plots*: empty points represent observations with no measurement errors; solid points represent those generated with error covariance Λ. Clusters are identified by different shapes. *Left*: clustering result produced by MCLUST-ME. Dashed line represents classification boundary for error-free observations; dotted line represents boundary for those with error covariance matrix Λ; solid line represents boundary produced by MCLUST. *Right*: clustering result produced by MCLUST. Solid line is the same as in the left plot.

Classification uncertainty In Figure 2, we visualize the classification uncertainty of each point produced by both methods. Observe that for MCLUST, highly uncertain points are found close to the decision boundary, regardless of error. For MCLUST-ME, points with measurement errors (solid) near the outer boundary (dotted) in the overlapping region tend to have high clustering uncertainties. Likewise, error-free points (empty) near the inner boundary (dashed) tend to have high uncertainties. This is consistent with our statement in Section 2.6.2.

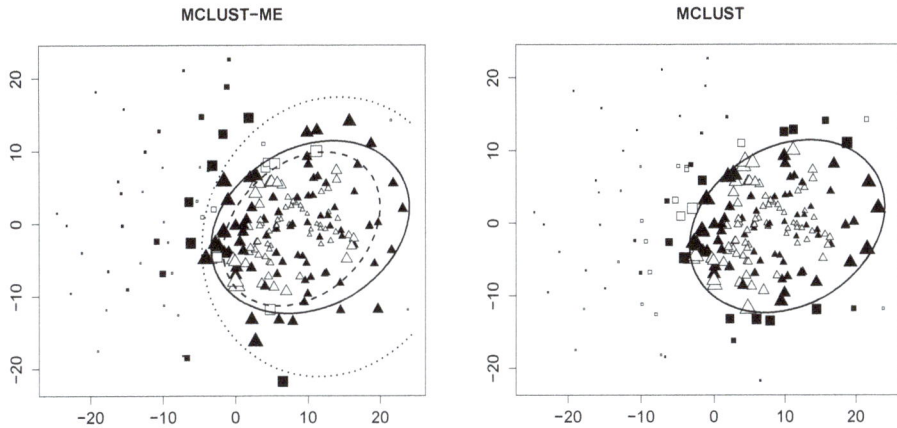

Figure 2. Clustering uncertainty of the sample generated with random seed = 7 and $\eta = 0.5$. Data points of larger size have a higher clustering uncertainty. All other graph attributes are the same as Figure 1.

Accuracy We first evaluate the performance of MCLUST and MCLUST-ME individually using ARI (between true group labels and predicted labels) as their performance measure. Figure 3 shows that for both methods, clustering accuracy tends to decrease as error proportion η increases. This is intuitively reasonable, because points associated with errors are more easily misclassified due to their high variability, and a larger proportion of such points means a lower overall accuracy.

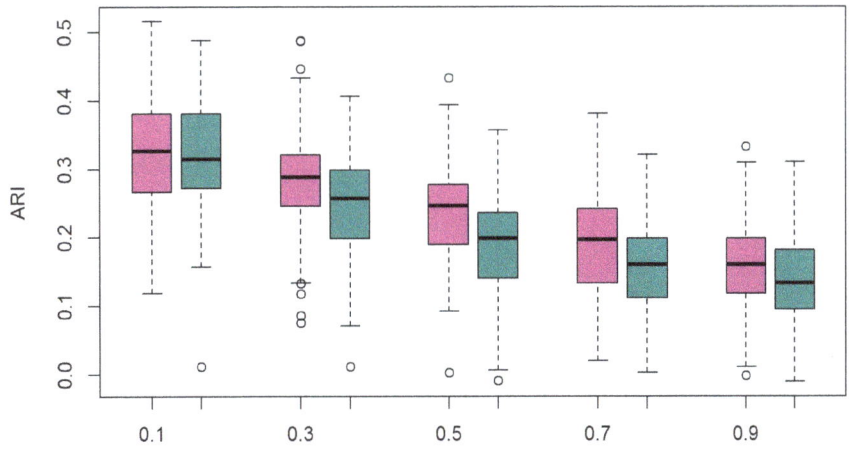

Figure 3. Adjusted Rand indices for MCLUST-ME and MCLUST. Five different proportions of erroneous observations (η) were considered. Magenta: MCLUST-ME; Dark Cyan: MCLUST.

Next, we compare the performances of MCLUST and MCLUST-ME by examining pairwise differences in ARI. Figure 4 shows that on average, MCLUST-ME has a slight advantage in accuracy, and it appears that this advantage is greatest when $\eta = 0.5$, and becomes smaller as η gets closer to either zero or one. In the latter situation, error covariances will tend to become constant (all equal to $36\mathbf{I}_2$ as $\eta \to 1$, or $\mathbf{0}$ as $\eta \to 0$) across all points, meaning that MCLUST-ME will behave more and more like MCLUST, hence diminishing MCLUST-ME's advantage in accuracy.

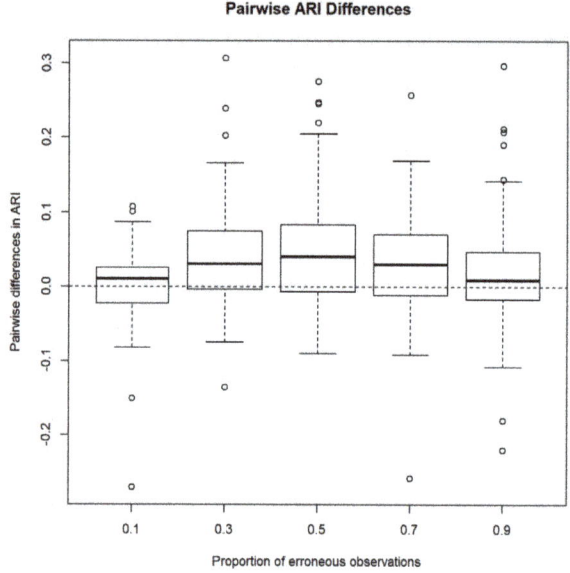

Figure 4. Pairwise difference in adjusted Rand indices between MCLUST-ME and MCLUST. Five different proportions of erroneous observations were considered.

Using a permutation test to test the hypotheses H_0 : $ARI_{MCLUST-ME} = ARI_{MCLUST}$ v.s. H_1 : $ARI_{MCLUST-ME} > ARI_{MCLUST}$, the p-values for the five cases are shown in Table 3. With the exception of $\eta = 0.1$, MCLUST-ME produced a significantly higher ARI than MCLUST.

Table 3. Permutation p-values for comparing MCLUST and MCLUST-ME ARI's.

η	0.1	0.3	0.5	0.7	0.9
p-value	0.256	0	0	0	0.002

Taking a closer look at the pairwise comparison when $\eta = 0.5$, Figure 5 shows that when MCLUST's accuracy is low, MCLUST-ME outperforms MCLUST most of the time, and when MCLUST's accuracy is relatively high, the two methods are less distinguishable on average.

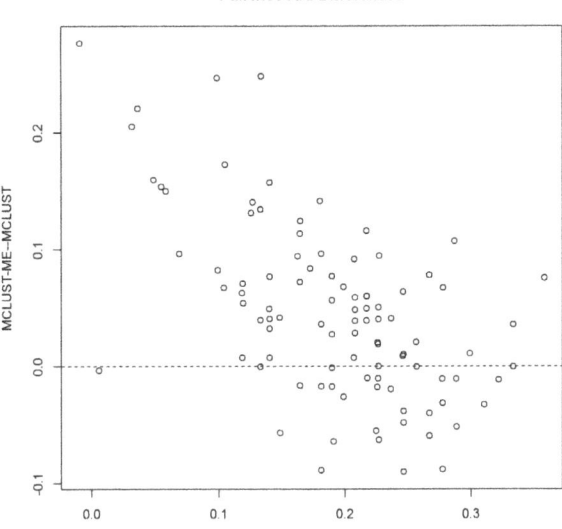

Figure 5. Pairwise difference in accuracy relative to MCLUST accuracy. *X-axis*: MCLUST ARI; *Y-axis*: Pairwise difference between MCLUST-ME and MCLUST ARI values.

3.2. Simulation 2: Clustering Uncertainties and Magnitudes of Error Covariances

In this simulation, our focus is on investigating how clustering uncertainties differ between MCLUST-ME and MCLUST: in particular, we want to see how the magnitudes of error covariances affect the uncertainty estimates. For this purpose, we will let the magnitudes of error covariances vary in a wide range.

3.2.1. Data Generation

The data were generated from a two-component bivariate normal mixture distribution with errors whose magnitudes are uniformly distributed. The data generation process is as follows.

(1) Generate $\{S_i\}_{i=1}^n$ i.i.d. from Uniform$(0, S)$, where S_i denotes the magnitude of error covariance for observation i.
(2) Generate $\{z_i\}_{i=1}^n$ i.i.d. from Bernoulli(τ). Parameter τ will be the mixing proportion.
(3) For $i = 1, ..., n$, generate y_i from

$$z_i N_2(\mu_1, \Sigma_1 + S_i I_2) + (1 - z_i) N_2(\mu_2, \Sigma_2 + S_i I_2),$$

where I_2 denotes the 2-dimensional identity matrix.

The parameter values are set as follows; $\mu_1 = (-10, 0)^T$, $\mu_2 = (10, 0)^T$, $\Sigma_1 = \Sigma_2 = 100 I_2$, $n = 200$, $\tau_1 = \tau_2 = 0.5$, and $S = 100$. We chose these parameter values so that there will be quite many points near the classification boundary: these points tend to have high classification uncertainties. We want to see, under MCLUST-ME and under MCLUST, how the error magnitudes, S_i, will affect the estimated classification uncertainties (defined in (14)) of the points.

3.2.2. Simulation Procedure

The simulation proceeds as follows.

(1) Generate a random sample following Section 3.2.1.
(2) Run MCLUST and MCLUST-ME, fixing $G = 2$. Initiate with true memberships.
(3) Record cluster membership probabilities and MLEs for model parameters upon convergence.

3.2.3. Simulation 2 Results

In Figure 6, we show the clustering results from MCLUST-ME and MCLUST ($G = 2$). On this data set, the hard partitioning results do not differ much between the two methods: only two points were classified differently by the two methods (highlighted by black circles).

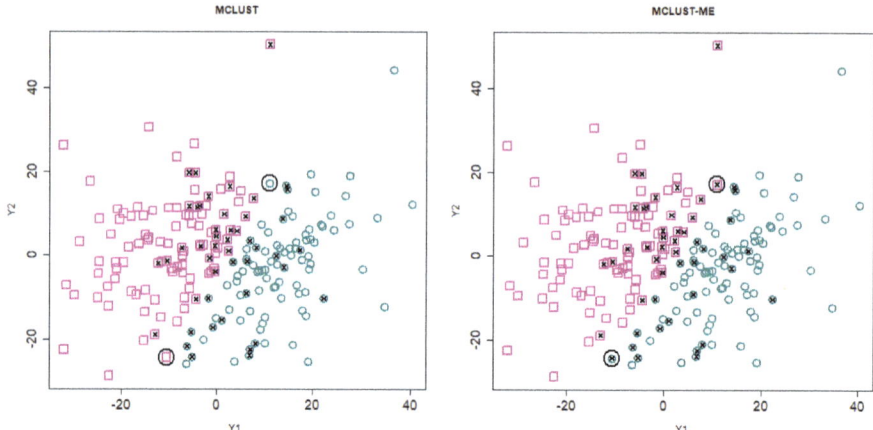

Figure 6. Clustering results for Simulation 2. The clustering results are indicated by different colors and symbols. Points with crosses are misclassified points. The two points that are classified differently by MCLUST-ME and MCLUST are circled in black.

Our focus here is on comparing the classification uncertainties estimated under the two methods. For MCLUST, the uncertainty measure for a point depends only on the point location and estimated centers and covariance matrices of the two clusters. Under MCLUST-ME, the uncertainty measure will also depend on the error covariance associated with the point. When two points are at the same location, MCLUST-ME will give higher uncertainty estimate to the point with greater error covariances (see Equation (12)), which is reasonable. In Figure 7, for each observation, we visualize the change in estimated membership probability to cluster 1 between MCLUST and MCLUST-ME with respect to the magnitude of its error covariance(S_i): the closer the membership probability is to 0.5 the higher the classification uncertainty. The points with most changes in estimated membership probabilities are highlighted in Figure 8. Relative to MCLUST, the MCLUST-ME model tends to adjust the classification uncertainties upwards for points with high error covariances and downwards for points with low error covariances. In other words, relative to the MCLUST-ME results, MCLUST tends to overestimate clustering uncertainties for points with low error covariances and underestimate clustering uncertainties for points with high error covariances. This is expected, as MCLUST treats all points as measured with no errors and absorbs all individual measurement/estimation errors into the variance estimates for the two clusters. As a crude approximation, one can think that MCLUST effectively treats each point as having an error covariance matrix close to the average of all true error covariances. However, the up or down changes in membership probabilities (and thus uncertainty estimates) are not a simple function of S_i, and we do not see a clear-cut boundary between the ups and downs in Figure 7, as the estimates of membership probabilities are also affected by differences in estimates of centers and covariance matrices of the two clusters.

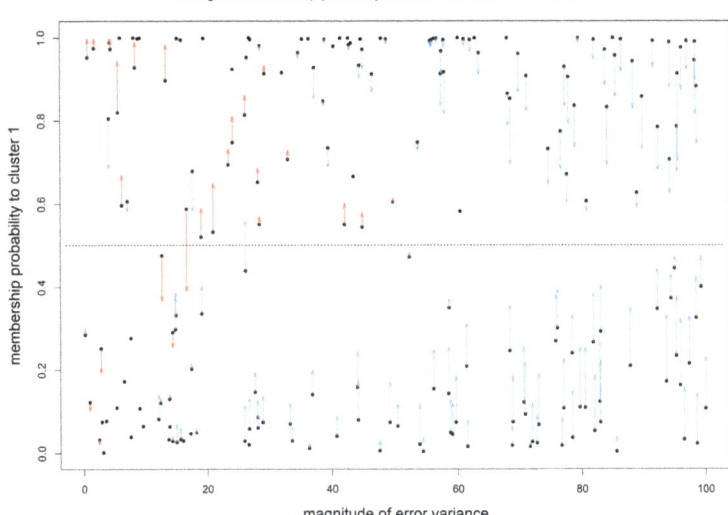

Figure 7. Change in estimated membership probability to cluster 1 from MCLUST to MCLUST-ME, plotted against error magnitude. *X-axis*: magnitude of error covariance, S_i; *Y-axis*: estimated membership probabilities to cluster 1 by MCLUST (black dots) and by MCLUST-ME (arrowheads). Changes in estimated membership probabilities from MCLUST to MCLUST-ME are highlighted by arrows (no arrow indicates a change less than 0.01). Blue and red arrows indicate an increase and decrease in estimated clustering uncertainty, respectively. With two clusters, the closer the estimated membership probability to 0.5, the higher the classification uncertainty; the classification membership changes when an arrow crosses the horizontal line at 0.5 (the dashed line).

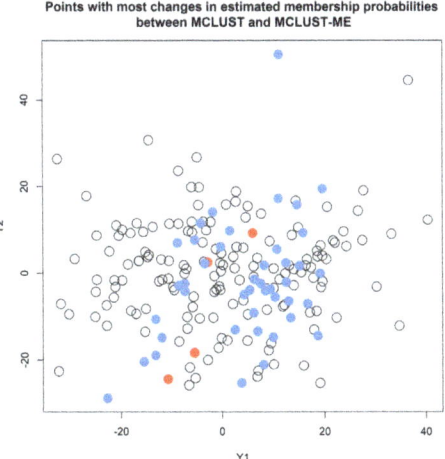

Figure 8. Points with most changes in estimated membership probabilities to cluster 1 from MCLUST to MCLUST-ME. The colored dots correspond to points with a change greater than 0.1 in estimated membership probability to cluster 1. Blue and red colors indicate an increase and decrease in estimated clustering uncertainty, respectively.

3.3. A Real Data Example

3.3.1. Data Description

The data come from an unpublished study on the model plant *Arabidopsis thaliana*. Researchers employed RNA-Seq to create a temporal profiling of *Arabidopsis* transcriptome over a *12h* period, with the aim of investigating plant innate immunity after elicitation of leaf tissue with flg22—a 22-amino-acid epitope of bacterial flagellin. A total of 33 *A. thaliana* Col-0 plants were grown in a controlled environment. Fifteen were treated with flg22, 15 with water, and the other 3 were left untreated. At each of five time points (*10 min, 1 h, 3 h, 6 h, 12 h*), three flg22-treated and three water-treated plants were harvested and prepared for RNA-Seq analysis.

A negative binomial regression model was fitted to each row (i.e., each gene) of the RNA-Seq count data. The regression model was parameterized such that the first five regression coefficients correspond to log fold changes in mean relative expression level between flg22- and water-treated groups at the five time points, which make up the temporal profile of each gene. The regression coefficients were estimated by the MLEs using the R package NBPSeq [24]. Furthermore, based on asymptotic normality of MLE, the covariance matrix of the log fold changes can be estimated by inverting the observed information matrix. For the current study, we will use the estimated regression coefficients and associated variance–covariance matrices for a subset of 1000 randomly selected genes at two of the time points (*1 h* and *3 h*) as input for the clustering analysis, as the gene expressions are most active at these two time points.

3.3.2. Cluster Analysis

We applied MCLUST-ME and MCLUST to the data. Both methods have their highest BIC values when $G = 2, 3,$ or 4. We focus on the $G = 2$ results as it is simple and yet illuminates the key differences between the two methods. In Figure 9, we show the clustering results from the two methods. In this example, both clustering methods show one cluster near the center and another cluster wrapping around it. This makes sense in the context of a gene expression study: the center cluster roughly represent genes that are not differentially expressed (non-DE) at these two time points; the outer cluster roughly represent genes that are differentially expressed (DE). In Figure 9, we see one signature difference between the two clustering methods: MCLUST gives a smooth boundary, whereas in the MCLUST-ME results, the two clusters are interspersed. This is expected from our theoretical analysis earlier and consistent with Simulation 1 results.

Table 4 summarizes the number of points that are classified differently by the two methods. In Figure 10, we show the standard errors (square roots of the diagonal entries of the error covariance) of the log fold changes estimated at 1 h and 3 h, with points classified differently by the two methods highlighted in colors. When we look at the points that are clustered differently by the two methods, we noticed that they tend to be the points either with very low or very high error covariances (relative to the average error covariance). This is expected as we understand that MCLUST absorbs all the individual error covariances into the estimation of the covariances of the two clusters, and thus is effectively using a middle-of-the-pack error covariance to treat each point. Therefore, we expect the differences in clustering results tend to show up among points with either very high or very low error covariances. This observation is also consistent with what we see in Simulation 2.

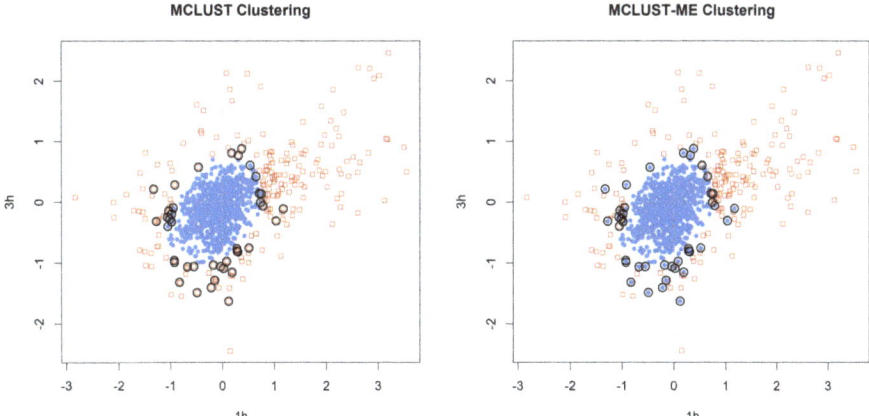

Figure 9. Clustering analysis of the log fold changes of 1000 genes randomly selected from the *Arabidopsis* data set. Two-group clustering of the data with MCLUST-ME and MCLUST, showing log fold changes at 1 h and 3 h. Groups are distinguished by point shapes and colors, and identified as non-DE group (blue circles) and DE group (red squares). Observations classified differently by the two methods are circled in black.

Table 4. Contingency table for group labels predicted by MCLUST-ME and MCLUST.

	MCLUST-ME	
MCLUST	Non-DE	DE
Non-DE	775	10
DE	30	185

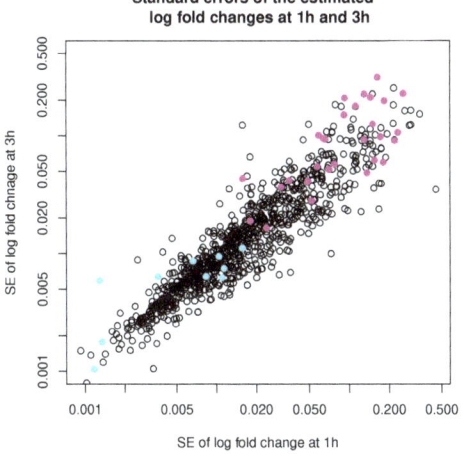

Figure 10. Standard errors of estimated log fold changes at 1 h and 3 h. Observations that are classified differently by MCLUST-ME and MCLUST are highlighted in colors. Magenta: classified as "DE" by MCLUST and as "non-DE" by MCLUST-ME; Cyan: classified as "non-DE" by MCLUST and as "DE" by MCLUST-ME. (Note that the axes are on the log scale.)

More interestingly, in this example, we see that the points (genes) that are classified into the "DE" cluster by MCLUST, but into the "non-DE" cluster by MCLUST-ME, tend to have high error covariances. In the MCLUST results, the clustering membership is completely determined by the magnitude of the two regression coefficients, which represent log fold changes between two experimental conditions at the two time points. In MCLUST-ME, membership calculation also considers the estimation uncertainty of the log fold changes. For gene expression data, we know that the uncertainty in log fold change estimation varies greatly (e.g., often depends on the mean expression levels). Although this example is a real data set with no ground truth on each point's actual group membership, it seems reasonable that points with moderate log fold changes but high error variances should be classified into the non-DE cluster, as MCLUST-ME has done in our example. At the minimum, the MCLUST-ME results warn us that not all points with the same log fold changes are created equal, which is exactly the point we want to highlight in this article. Actually, this example is the data set that motivated us to consider incorporating uncertainty information into the clustering algorithm. In this example, explicitly modeling the error covariances clearly shows a difference.

The error covariance matrices were estimated, and thus associated with their own estimation errors. To get a sense of the uncertainty associated with estimating the error covariance matrices, we simulated additional sets of error covariance estimates by parametric bootstrapping: simulating copies of the RNA-seq data set based on parameters estimated from the real data set and estimating error covariance matrices from the simulated data sets. In Figure 11, we compare the square roots of the diagonal entries of two sets of simulated error covariance estimates (which correspond to the standard errors of the log fold changes at the two time points). We then tried MCLUST-ME method on the original data set with the two sets of simulated error covariance estimates: eight observations were classified differently due to the differences in error covariance estimates (see Table 5 for a summary). For a closer look, in Figure 12, we show the differences in the estimated membership probabilities (to the non-DE cluster) between the two runs of MCLUST-ME with different simulated error covariance estimates, and these differences were much less than the differences between the original MCLUST-ME and MCLUST results. These results show that the uncertainty in covariance estimation does lead to variation in the clustering results, but the variation is much less as compared to the differences between whether or not to model the estimation errors. In this sense, the MCLUST-ME method is robust to the uncertainty in the covariance estimation to a certain degree.

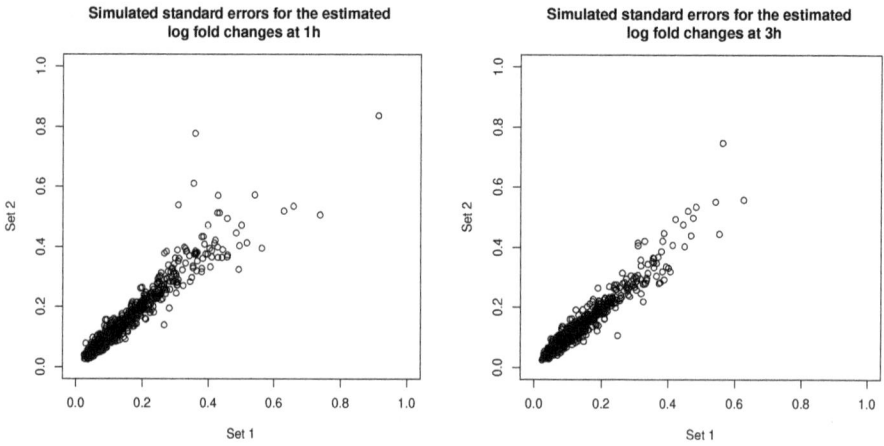

Figure 11. Comparing two sets of simulated standard errors for the estimated log fold changes at 1 h (**left**) and at 3 h (**right**). The standard errors correspond to the square roots of the diagonal entries of the simulated error covariance estimates.

Table 5. Contingency table for group labels predicted by MCLUST-ME with two sets of simulated error covariance estimates

	MCLUST-ME Run 1	
MCLUST-ME Run 2	Non-DE	DE
Non-DE	801	2
DE	6	191

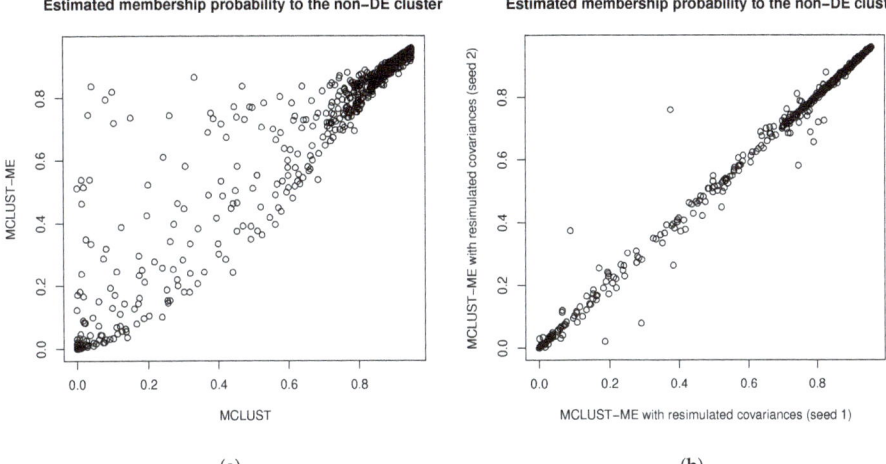

Figure 12. (**a**) Comparing membership probabilities to the "non-DE" cluster estimated by MCLUST-ME and by MCLUST. (**b**) Comparing membership probabilities to the "non-DE" cluster estimated by MCLUST-ME with two sets of simulated covariance estimates. The decision whether or not to model the error covariances will result in drastic changes in the estimated membership probabilities. In comparison, the uncertainties in covariance estimation cause much less changes in the estimated membership probabilities.

3.3.3. Comparison to kError

In Section 2.7, we reviewed the clustering method by the authors of [18], which models the error covariances of individual observations as in MCLUST-ME, but lacks the model-based components ($N_d(\mathbf{0}, \mathbf{\Sigma}_k)$) for modeling individual clusters. We implemented the kError algorithm according to the description in [18] and applied it to the RNA-Seq data set that we analyzed in the previous subsection, using the estimation error covariances as $\tilde{\mathbf{\Lambda}}_i$ and using the memberships predicted by MCLUST-ME as initial values. The clustering results by kError are shown in Figure 13, which can be compared with the MCLUST and MCLUST-ME results in Figure 9.

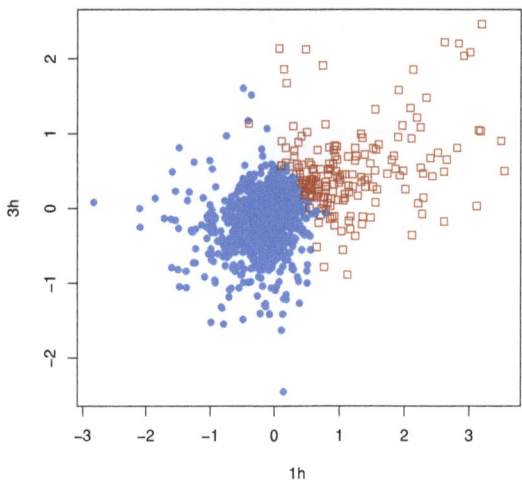

Figure 13. Two-grop clustering results by kError. We applied the kError method to the same RNA-Seq data set that was analyzed by MCLUST and MCLUST-ME. Compare with Figure 9.

For this data set, the two clusters estimated by MCLUST or MCLUST-ME have quite different Σ_k values: the covariance of the DE cluster is much greater in magnitude than that of the non-DE cluster. The DE cluster is enclosed by the non-DE cluster. Such a structure between the two clusters is difficult for kError method to capture. The way kError split the data sets into two clusters is similar to an ordinary k-means method. Interestingly, the two clusters by kError are interspersed without a clean-cut boundary and points with similar values but different covariances can belong to different clusters: This feature is similar to MCLUST-ME.

4. Conclusions and Discussion

In this paper, we proposed an extension to model-based clustering approach that accounts for known or estimated error covariances for data observed with uncertainty. The error covariances can often be estimated for data consisting of summary statistics, such as the regression coefficients from a regression analysis. We extended the EM algorithm implemented in MCLUST and implemented our new method MCLUST-ME in R [25].

A distinctive feature of MCLUST-ME is that the classification boundary separating the clusters is not always shared by all observations; instead, each distinct value of error covariance matrix corresponds to a different boundary. Using both simulated and a real data example, we have shown that under certain circumstances, explicitly accounting for estimation error distributions does lead to improved clustering results or new insights, where the degree of improvement depends on the distribution of error covariances.

It is not our intention to claim that MCLUST-ME is universally better than the original MCLUST. We are actually more interested in understanding when it will give different results than MCLUST: in other words, when it is beneficial to explicitly model the measurement error structures when performing clustering analysis. When covariances of estimation errors are roughly constant or small relative to the covariances of the clusters, MCLUST and MCLUST-ME yield highly similar results. We will tend to see meaningful differences when there is significant overlap among clusters (i.e., the difficult cases) and when there is a large variation in the magnitude of error variance.

There are a few natural extensions that can be implemented. For example, in this paper, we focused on the case where the variance–covariance matrices of the clusters are unconstrained (what MCLUST calls "VVV" type). One important feature of the original MCLUST method is that it allows structured constraints on the cluster variance–covariance matrices. Such extension is possible for MCLUST-ME. The main challenge for our current implementation of MCLUST-ME is computational. With MCLUST-ME, each point has its own error covariance matrix, and therefore we no longer have closed-form solutions for estimating the model parameters and have to rely on optimization routines. These factors make MCLUST-ME slower than the MCLUST implementation, but for reasonably-sized low-dimensional data sets, it is still manageable. The running time of the algorithm will depend on the number of clusters (G) and the size and dimension of the observed data. For our real data example, when we classify the 1000 two-dimensional data points into two clusters, it took 19 min. It took 23 h to classify the same data sets into six clusters (on a laptop workstation with an Xeon X3430 processor). To this end, improving the computation routine or exploring approximation methods is a future research topic.

The data and R code for reproducing the results in this paper is available online at https://github.com/diystat/MCLUST-ME-Genes.

Author Contributions: Conceptualization, Y.D.; data curation, W.Z.; formal analysis, W.Z. and Y.D.; investigation, W.Z. and Y.D.; methodology, W.Z. and Y.D.; software, W.Z. and Y.D.; supervision, Y.D.; validation, W.Z. and Y.D.; visualization, W.Z.; writing—original draft, W.Z. and Y.D.; writing—review & editing, W.Z. and Yanming Di. All authors have read and agreed to the published version of the manuscript.

Funding: This research was partially funded by the National Institute of General Medical Sciences of the National Institutes of Health under Award Number R01 GM104977.

Acknowledgments: The authors gratefully acknowledge Sarah Emerson, Duo Jiang, and Bin Zhuo for their valuable insight and comments, and Gitta Coaker for providing the RNA-seq experiment data. We would like to thank Joe Defilippis for IT support. We would like to thank both reviewers for their constructive comments that improved the paper.

Conflicts of Interest: The authors declare no conflicts of interest.

Abbreviations

The following abbreviations are used in this manuscript:

ARI	Adjusted Rand index
BIC	Bayesain information criterion
DE	Differentially expressed
EM	Expectation-maximization
MLE	Maximum likelihood estimate(s)
iid	independent and identically distributed
RI	Rand index

References

1. Fraley, C.; Raftery, A.E. Model-based clustering, discriminant analysis, and density estimation. *J. Am. Stat. Assoc.* **2002**, *97*, 611–631. [CrossRef]
2. Bouveyron, C.; Celeux, G.; Murphy, T.B.; Raftery, A.E. *Model-Based Clustering and Classification for Data Science: With Applications in R*; Cambridge University Press: Cambridge, UK, 2019; Volume 50.
3. Wolfe, J.H. Pattern clustering by multivariate mixture analysis. *Multivar. Behav. Res.* **1970**, *5*, 329–350. [CrossRef] [PubMed]
4. Fraley, C.; Raftery, A.E. Enhanced model-based clustering, density estimation, and discriminant analysis software: MCLUST. *J. Classif.* **2003**, *20*, 263–286. [CrossRef]
5. Fraley, C.; Raftery, A.E.; Murphy, T.B.; Scrucca, L. *Mclust Version 4 for R: Normal Mixture Modeling for Model-Based Clustering, Classification, and Density Estimation*; Tech. Rep. No. 597; Department of Statistics, University of Washington: Washington, DC, USA, 2012.

6. Scrucca, L.; Fop, M.; Murphy, T.B.; Raftery, A.E. mclust 5: Clustering, classification and density estimation using Gaussian finite mixture models. *R J.* **2016**, *8*, 289. [CrossRef] [PubMed]
7. Banfield, J.D.; Raftery, A.E. Model-based Gaussian and non-Gaussian clustering. *Biometrics* **1993**, *49*, 803–821. [CrossRef]
8. Dempster, A.P.; Laird, N.M.; Rubin, D.B. Maximum likelihood from incomplete data via the EM algorithm. *J. R. Stat. Society. Ser. B (Methodological)* **1977**, *39*, 1–38.
9. Celeux, G.; Govaert, G. Gaussian parsimonious clustering models. *Pattern Recognit.* **1995**, *28*, 781–793. [CrossRef]
10. Schwarz, G. Estimating the dimension of a model. *Ann. Stat.* **1978**, *6*, 461–464. [CrossRef]
11. Dasgupta, A.; Raftery, A.E. Detecting features in spatial point processes with clutter via model-based clustering. *J. Am. Stat. Assoc.* **1998**, *93*, 294–302. [CrossRef]
12. Zhang, W. Model-based Clustering Methods in Exploratory Analysis of RNA-Seq Experiments. Ph.D. Thesis, Oregon State University, Corvallis, OR, USA, 2017.
13. Dwyer, P.S. Some applications of matrix derivatives in multivariate analysis. *J. Am. Stat. Assoc.* **1967**, *62*, 607–625. [CrossRef]
14. Byrd, R.H.; Lu, P.; Nocedal, J.; Zhu, C. A limited memory algorithm for bound constrained optimization. *SIAM J. Sci. Comput.* **1995**, *16*, 1190–1208. [CrossRef]
15. Murtagh, F.; Raftery, A.E. Fitting straight lines to point patterns. *Pattern Recognit.* **1984**, *17*, 479–483. [CrossRef]
16. Fraley, C. Algorithms for model-based Gaussian hierarchical clustering. *SIAM J. Sci. Comput.* **1998**, *20*, 270-281. [CrossRef]
17. Karlis, D.; Xekalaki, E. Choosing initial values for the EM algorithm for finite mixtures. *Comput. Stat. Data Anal.* **2002**, *41*, 577–900. [CrossRef]
18. Kumar, M.; Patel, N.R. Clustering data with measurement errors. *Comput. Stat. Data Anal.* **2007**, *51*, 6084–6101. [CrossRef]
19. Tjaden, B. An approach for clustering gene expression data with error information. *BMC Bioinform.* **2006**, *7*, 17. [CrossRef] [PubMed]
20. Rand, W.M. Objective Criteria for the Evaluation of Clustering Methods. *J. Am. Stat. Assoc.* **1971**, *66*, 846–850. [CrossRef]
21. Santos, J.M.; Embrechts, M. On the use of the adjusted Rand index as a metric for evaluating supervised classification. In Proceedings of the ICANN (International Conference on Artificial Neural Networks), Limassol, Cyprus, 14–17 September 2009; Springer: Berlin, Germany, 2009; pp. 175–184.
22. Hubert, L.; Arabie, P. Comparing partitions. *J. Classif.* **1985**, *2*, 193–218. [CrossRef]
23. Qannari, E.M.; Courcoux, P.; Faye, P. Significance test of the adjusted Rand index. Application to the free sorting task. *Food Qual. Prefer.* **2014**, *32*, 93–97. [CrossRef]
24. Di, Y. Single-gene negative binomial regression models for RNA-Seq data with higher-order asymptotic inference. *Stat. Its Interface* **2015**, *8*, 405. [CrossRef] [PubMed]
25. R Core Team. *R: A Language and Environment for Statistical Computing*; R Foundation for Statistical Computing: Vienna, Austria, 2019.

© 2020 by the authors. Licensee MDPI, Basel, Switzerland. This article is an open access article distributed under the terms and conditions of the Creative Commons Attribution (CC BY) license (http://creativecommons.org/licenses/by/4.0/).

Article
A Pathway-Based Kernel Boosting Method for Sample Classification Using Genomic Data

Li Zeng [1], Zhaolong Yu [2] and Hongyu Zhao [1,2,*]

[1] Department of Biostatistics, Yale University, New Haven, CT 06511, USA
[2] Interdepartmental Program in Computational Biology and Bioinformatics, Yale University, New Haven, CT 06511, USA
* Correspondence: hongyu.zhao@yale.edu; Tel.: +1-2037853613

Received: 12 August 2019; Accepted: 28 August 2019; Published: 31 August 2019

Abstract: The analysis of cancer genomic data has long suffered "the curse of dimensionality." Sample sizes for most cancer genomic studies are a few hundreds at most while there are tens of thousands of genomic features studied. Various methods have been proposed to leverage prior biological knowledge, such as pathways, to more effectively analyze cancer genomic data. Most of the methods focus on testing marginal significance of the associations between pathways and clinical phenotypes. They can identify informative pathways but do not involve predictive modeling. In this article, we propose a Pathway-based Kernel Boosting (PKB) method for integrating gene pathway information for sample classification, where we use kernel functions calculated from each pathway as base learners and learn the weights through iterative optimization of the classification loss function. We apply PKB and several competing methods to three cancer studies with pathological and clinical information, including tumor grade, stage, tumor sites and metastasis status. Our results show that PKB outperforms other methods and identifies pathways relevant to the outcome variables.

Keywords: classification; gene set enrichment analysis; boosting; kernel method

1. Introduction

High-throughput genomic technologies have enabled cancer researchers to study the associations between genes and clinical phenotypes of interest. A large number of cancer genomic data sets have been collected with both genomic and clinical information from the patients. The analyses of these data have yielded valuable insights on cancer mechanisms, subtypes, prognosis and treatment response.

Although many methods have been developed to identify genes informative of clinical phenotypes and build prediction models from these data, it is often difficult to interpret the results with single-gene focused approaches, as one gene is often involved in multiple biological processes and the results are not robust when the signals from individual genes are weak. As a result, pathway-based methods have gained much popularity (e.g., Subramanian et al. [1]). A pathway can be considered as a set of genes that are involved in the same biological process or molecular function. It has been shown that gene-gene interactions may have stronger effects on phenotypes when the genes belong to the same pathway or regulatory network [2]. There are many pathway databases available, such as the Kyoto Encyclopedia of Genes and Genomes [3] (KEGG), the Pathway Interaction Database [4] and Biocarta [5]. By utilizing pathway information, researchers may aggregate weak signals from the same pathway to identify relevant pathways with better power and interpretability. Many pathway-based methods, such as GSEA [1], LSKM [6] and SKAT [7], focus on testing the significance of pathways. These methods consider each pathway separately and evaluate statistical significance for its relevance to the phenotype. In other words, these methods study each pathway separately without considering the effects of other pathways.

Given that many pathways likely contribute to the onset and progression of a disease [8–10]. It is of interest to study the contribution of a specific pathway to phenotypes conditional on the effects of other pathways. This is usually achieved by regression models. Wei and Li [11] and Luan and Li [12] proposed two similar models, Nonparametric Pathway-based Regression (NPR) and Group Additive Regression (GAR). Both models employ a boosting framework, construct base learners from individual pathways and perform prediction through additive models. Due to the additivity at the pathway level, these models only considered interactions among genes within the same pathway but not across pathways. Since our proposed method is motivated by the above two models, more details of these models will be described in Section 2. In genomics data analysis, multiple kernel methods [13,14] are also commonly used when predictors have group structures. In these methods, one kernel is assigned to each group of predictors and a meta-kernel is computed as a weighted sum of the individual kernels. The kernel weights are estimated through optimization and can be considered as a measure of pathway importance. Multiple kernel methods have been used to integrate multi-pathway information or multi-omics data sets and have achieved state-of-the-art performance in predictions of various outcomes [15–17].

In this paper, we propose a Pathway-based Kernel Boosting (PKB) method for sample classification. In our boosting framework, we use the second order approximation of the loss function instead of the first order approximation used in the usual gradient descent boosting method, which allows for deeper descent at each step. We introduce two types of regularizations (L_1 and L_2) for selection of base learners in each iteration and propose algorithms for solving the regularized problems. In Section 3.1, we conduct simulation studies to evaluate the performance of PKB, along with four other competing methods. In Section 3.2, we apply PKB to three cancer genomics data sets, where we use gene expression data to predict several patient phenotypes, including tumor grade, stage, tumor site and metastasis status.

2. Materials and Methods

Suppose our observed data are collected from N subjects. For subject i, we use a p dimensional vector $\mathbf{x}_i = (x_{i1}, x_{i2}, \ldots, x_{ip})$ to denote the normalized gene expression profile and $y_i \in \{1, -1\}$ to denote its class label. Similarly, the gene expression levels of a given pathway m with p_m genes can be represented by $\mathbf{x}_i^{(m)} = (\mathbf{x}_{i1}^{(m)}, \mathbf{x}_{i2}^{(m)}, \ldots, \mathbf{x}_{ip_m}^{(m)})$, which is a sub-vector of $\mathbf{x}_i$.

The log loss function is commonly used in binary classifications with the following form:

$$l(y, F(\mathbf{x})) = \log(1 + e^{-yF(\mathbf{x})}),$$

and is minimized by

$$F^*(\mathbf{x}) = \log \frac{p(y=1|\mathbf{x})}{p(y=-1|\mathbf{x})},$$

which is exactly the log odds function. Thus the sign of an estimated $F(\mathbf{x})$ can be used to classify sample $\mathbf{x}$ as 1 or -1. Since genes within the same pathway likely have much stronger interactions than genes in different pathways, in our pathway-based model setting, we assume additive effects across pathways and focus on capturing gene interactions within pathways:

$$F(\mathbf{x}) = \sum_{m=1}^{M} H_m(\mathbf{x}^{(m)}),$$

where each H_m is a nonlinear function that only depends on the expression levels of genes in the mth pathway and summarizes its contribution to the log odds function. Due to the additive nature of this model, it only captures gene interactions within each pathway but not across pathways.

Two existing methods, NPR [11] and GAR [12], employed the Gradient Descent Boosting (GDB) framework [18] to estimate the functional form of $F(\mathbf{x})$ nonparametrically. GDB can be considered as a functional gradient descent algorithm to minimize the empirical loss function, where in each descent iteration, an increment function that best aligns with the negative gradient of the loss function (evaluated at each sample point) is selected from a space of base learners and then added to the target function $F(\mathbf{x})$. NPR and GAR extended GDB to be pathway-based by applying the descent step to each pathway separately and selecting the base learner from the pathway that provides the best fit to the negative gradient.

NPR and GAR differ in how they construct base learners from each pathway: NPR uses regression trees and GAR uses linear models. Due to the linearity assumption of GAR, it lacks the ability to capture complex interactions among genes in the same pathway. Using regression tree as base learners enables NPR to model interactions, however, there is no regularization in the gradient descent step, which can lead to selection bias that prefers larger pathways.

Motivated by NPR and GAR, we propose the PKB model, where we employ kernel functions as base learners, optimize loss function with second order approximation [19] which gives Newton-like descent speed and also incorporates regularization in selection of pathways in each boosting iteration.

2.1. PKB Model

Kernel methods have been applied to a variety of statistical problems, including classification [20], regression [21], dimension reduction [22] and others. Results from theories of Reproducing Kernel Hilbert Space [23] have shown that kernel functions can capture complex interactions among features. For pathway m, we construct a kernel-based function space as the space for base learners

$$\mathcal{G}_m = \{g(\mathbf{x}) = \sum_{i=1}^{N} K_m(\mathbf{x}_i^{(m)}, \mathbf{x}^{(m)}) \beta_i + c : \beta_1, \beta_2, \ldots, \beta_N, c \in R\},$$

where $K_m(\cdot, \cdot)$ is a kernel function that defines similarity between two samples only using genes in the mth pathway. The overall base learner space is the union of the spaces constructed from each pathway alone: $\mathcal{G} = \cup_{m=1}^{M} \mathcal{G}_m$.

Estimation of the target function $F(\mathbf{x})$ is obtained through iterative minimization of the empirical loss function evaluated at the observed data. The empirical loss is defined as

$$L(\mathbf{y}, \mathbf{F}) = \frac{1}{N} \sum_{i=1}^{N} l(y_i, F(\mathbf{x}_i)),$$

where $\mathbf{F} = (F(\mathbf{x}_1), F(\mathbf{x}_2), \ldots, F(\mathbf{x}_N))$. In the rest of this article, we will use the bold font of a function to represent the vector of the function evaluated at the observed $\mathbf{x}_i$'s. Assume that at iteration t, the estimated target function is $F_t(\mathbf{x})$. In the next iteration, we aim to find the best increment function $f \in \mathcal{G}$ and add it to $F_t(\mathbf{x})$. Expanding the empirical loss at $\mathbf{F}_t$ to the second order, we can get the following approximation

$$L_{\text{approx}}(\mathbf{y}, \mathbf{F}_t + \mathbf{f}) = L(\mathbf{y}, \mathbf{F}_t) + \frac{1}{N} \sum_{i=1}^{N} [h_{t,i} f(\mathbf{x}_i) + \frac{1}{2} q_{t,i} f(\mathbf{x}_i)^2], \tag{1}$$

where

$$h_{t,i} = \frac{\partial L(\mathbf{y}, \mathbf{F}_t)}{\partial F_t(\mathbf{x}_i)} = -\frac{y_i}{1 + e^{y_i F_t(\mathbf{x}_i)}},$$

$$q_{t,i} = \frac{\partial^2 L(\mathbf{y}, \mathbf{F}_t)}{\partial F_t(\mathbf{x}_i)^2} = \frac{e^{y_i F_t(\mathbf{x}_i)}}{(1 + e^{y_i F_t(\mathbf{x}_i)})^2}$$

are the first order and second order derivatives with respect to each $F_t(\mathbf{x}_i)$, respectively. We propose a regularized loss function that incorporates both the approximated loss and a penalty on the complexity of f:

$$L_R(\mathbf{f}) = L_{\text{approx}}(\mathbf{y}, \mathbf{F}_t + \mathbf{f}) + \Omega(f) \qquad (2)$$

$$= \frac{1}{N}\sum_{i=1}^{N}\frac{q_{i,t}}{2}\left(\frac{h_{i,t}}{q_{i,t}} + f(\mathbf{x}_i)\right)^2 + \Omega(f) + C(\mathbf{y}, \mathbf{F}_t), \qquad (3)$$

where $\Omega(\cdot)$ is the penalty function. Since $f \in \mathcal{G}$ is a linear combination of kernel functions calculated from a specific pathway, the norm of the combination coefficients can be used to define $\Omega(\cdot)$. We consider both L_1 and L_2 norm penalties and solutions regarding each penalty option are presented in Sections 2.1.1 and 2.1.2, respectively. $C(\mathbf{y}, \mathbf{F}_t)$ is a constant term with respect to f. Therefore, we only use the first two terms of Equation (3) as the working loss function in our algorithms. We will also drop $C(\mathbf{y}, \mathbf{F}_t)$ in the expression of $L_R(\mathbf{f})$ in the following sections for brevity. Such a penalized boosting step has been employed in several methods (e.g., Johnson and Zhang [24]). Intuitively, the regularized loss function would prefer simple solutions that also fit the observed data well, which usually leads to better generalization capability to unseen data.

We then optimize the regularized loss for the best increment direction

$$\hat{f} = \arg\min_{f \in \mathcal{G}} L_R(\mathbf{f}).$$

Given the direction, we find the deepest descent step length by minimizing over the original loss function

$$\hat{d} = \arg\min_{d \in R^+} L(\mathbf{y}, \mathbf{F}_t + d\hat{f}),$$

and update the target function to $F_{t+1}(\mathbf{x}) = F_t(\mathbf{x}) + \nu d\hat{f}$, where ν is a learning rate parameter. The above fitting procedure is repeated until a certain pre-specified number of iterations is reached. The complete procedure of the PKB algorithm is shown in Table 1.

Table 1. An overview of the Pathway-based Kernel Boosting (PKB) algorithm.

---

1. Initialize target function as an optimal constant:

$$F_0(\mathbf{x}) = \arg\min_{r \in R} \frac{1}{n}\sum_{i=1}^{N} l(y_i, r)$$

For t from 0 to T-1 (maximum number of iterations) do:

2. calculate the first and second derivatives:

$$h_{t,i} = -\frac{y_i}{1 + e^{y_i F_t(\mathbf{x}_i)}}, q_{t,i} = \frac{e^{y_i F_t(\mathbf{x}_i)}}{(1 + e^{y_i F_t(\mathbf{x}_i)})^2}$$

3. optimize the regularized loss function in the base learner space:

$$\hat{f} = \arg\min_{f \in \mathcal{G}} L_R(\mathbf{f})$$

4. find the step length with the steepest descent:

$$\hat{d} = \arg\min_{d \in R^+} L(\mathbf{y}, \mathbf{F}_t + d\hat{f})$$

5. update the target function:

$$F_{t+1}(\mathbf{x}) = F_t(\mathbf{x}) + \nu d\hat{f}(\mathbf{x})$$

End For
return $F_T(\mathbf{x})$

2.1.1. L_1 Penalized Boosting

The core step of PKB is the optimization of the regularized loss function (see step 3 of Table 1). Note that $\mathcal{G}$ is the union of the pathway-based learner spaces, thus

$$\hat{f} = \arg\min_{\mathcal{G}} L_R(\mathbf{f})$$
$$= \arg\min_{\hat{f}_m} \{L_R(\hat{f}_m) : \hat{f}_m = \arg\min_{f \in \mathcal{G}_m} L_R(\mathbf{f}), m = 1, 2, \ldots, M\}.$$

To solve for $\hat{f}$, it is sufficient to obtain the optimal $\hat{f}_m$ in each pathway-based subspace. Due to the way we construct the subspaces, in a given pathway m, f takes a parametric form as a linear combination of the corresponding kernel functions. This helps us further reduce the optimization problem to

$$\min_{f \in \mathcal{G}_m} L_R(\mathbf{f}) = \min_{\beta,c} \frac{1}{N} \sum_{i=1}^{N} \frac{q_{i,t}}{2} \left(\frac{h_{i,t}}{q_{i,t}} + K_{m,i}^T \beta + c \right)^2 + \Omega(f) \quad (4)$$

$$= \min_{\beta,c} \frac{1}{N} (\eta_t + K_m \beta + 1_N c)^T W_t (\eta_t + K_m \beta + 1_N c) + \Omega(f), \quad (5)$$

where

$$\eta_t = \left(\frac{h_{1,t}}{q_{1,t}}, \frac{h_{2,t}}{q_{2,t}}, \ldots, \frac{h_{N,t}}{q_{N,t}} \right)^T,$$
$$W_t = \text{diag}\left(\frac{q_{1,t}}{2}, \frac{q_{2,t}}{2}, \ldots, \frac{q_{N,t}}{2} \right),$$
$$K_m = \left[K_m(\mathbf{x}_i^{(m)}, \mathbf{x}_j^{(m)}) \right]_{i,j=1,2,\ldots,N}.$$

$K_{m,i}$ is the ith column of kernel matrix K_m and 1_N is an N by 1 vector of 1's. We use the L_1 norm $\Omega(f) = \lambda \|\beta\|_1$, as the penalty term, where λ is a tuning parameter adjusting the amount of penalty we impose on model complexity. We also prove that after certain transformations, the optimization can be converted to a LASSO problem without intercept

$$\min_{\beta} \frac{1}{N} \|\tilde{\eta} + \tilde{K}_m \beta\|_2^2 + \lambda \|\beta\|_1, \quad (6)$$

where

$$\tilde{\eta} = W_t^{\frac{1}{2}} \left[I_N - \frac{1_N 1_N^T W_t}{\text{tr}(W_t)} \right] \eta_t$$
$$\tilde{K}_m = W_t^{\frac{1}{2}} \left[I_N - \frac{1_N 1_N^T W_t}{\text{tr}(W_t)} \right] K_m.$$

Therefore, β can be efficiently estimated using existing LASSO solvers. The proof of the equivalence between the two problems is provided in Section 1 of the Supplementary Materials.

2.1.2. L_2 Penalized Boosting

In the L_2 penalized boosting, we replace $\Omega(f)$ in the objective function of (5) with $\lambda\|\beta\|_2^2$. Following the same transformation as that in Section 2.1.1, the objective can also be converted to a standard Ridge Regression (see Section 1 of Supplementary Materials)

$$\min_{\beta} \frac{1}{N}\|\tilde{\eta} + \tilde{K}_m\beta\|_2^2 + \lambda\|\beta\|_2^2, \tag{7}$$

which allows closed form solution

$$\hat{\beta} = -(\tilde{K}_m^T\tilde{K}_m + N\lambda I_N)^{-1}\tilde{K}_m^T\tilde{\eta}.$$

Both the L_1 and L_2 boosting algorithms require the specification of the penalty parameter λ, which controls step length (the norm of fitted β) in each iteration and additionally controls solution sparsity in the L_1 case. Feasible choices of λ might be different for different scenarios, depending on the input data and also the choice of the kernel. Either too small or too large λ values would lead to big leaps or slow descent speed. Under the L_1 penalty, poor choices of λ can even result in all-zero β, which makes no change to the target function. Therefore, we also incorporate an optional automated procedure to choose the value of λ in PKB. Computational details of the procedure are provided in Section 2 of the Supplementary Materials. We recommend the use of the automated procedure to calculate a feasible λ and try a range of values around it (e.g., the calculated value multiplies $1/25, 1/5, 1, 5, 25$) for improved performance.

Lastly, the final target function at iteration T can be written as

$$F_T(\mathbf{x}) = \sum_{m=1}^{M}\sum_{i=1}^{N} K_m(\mathbf{x}_i^{(m)}, \mathbf{x}^{(m)})\beta_i^{(m)} + C,$$

where $\beta^{(m)} = (\beta_1^{(m)}, \beta_2^{(m)}, \ldots, \beta_N^{(m)})$ are the combination coefficients of kernel functions from pathway m. We use $\|\beta^{(m)}\|_2$ as a measure of importance (or weight) in the target function. It is obvious that only the pathways that are selected at least once in the boosting procedure will have non-zero weights. Because $F_T(\mathbf{x})$ is an estimation of the log odds function, $\text{sign}[F_T(\mathbf{x})]$ is used as the classification rule to assign $\mathbf{x}$ to 1 or -1.

3. Results

3.1. Simulation Studies

We use simulation studies to assess the performance of PKB. We consider the following three underlying true models:

- Model 1:
$$F(\mathbf{x}) = 2x_1^{(1)} + 3x_2^{(1)} + \exp(0.8x_1^{(2)} + 0.8x_2^{(2)}) + 4x_1^{(3)}x_2^{(3)}$$

- Model 2:
$$F(\mathbf{x}) = 4\sin(x_1^{(1)} + x_2^{(1)}) + 3|x_1^{(2)} - x_2^{(2)}| + 2x_1^{(3)^2} - 2x_2^{(3)^2}$$

- Model 3:
$$F(\mathbf{x}) = 2\sum_{m=1}^{10}\|\mathbf{x}^{(m)}\|_2$$

where $F(\mathbf{x})$ is the true log odds function and $x_i^{(m)}$ represents the expression level of the ith gene in the mth pathway. We include different functional forms of pathway effects in $F(\mathbf{x})$, including linear, exponential, polynomial and others. In models 1 and 2, only two genes in each of the first three pathways are informative to sample classes; in model 3, only genes in the first ten pathways are informative. We generated a total of six datasets, two for each model, with different numbers of irrelevant pathways (M = 50 and 150) corresponding to different noise levels. We set the size of pathways to 5 and sample size to 900 in all simulations. Gene expression data ($\mathbf{x}_i$'s) were generated following standard normal distribution. We then calculated the log odds $F(\mathbf{x}_i)$ for each sample and use the median-centered $F(\mathbf{x}_i)$ values to generate corresponding binary outcomes $y_i \in \{-1, 1\}$ (We usethe median-centered $F(\mathbf{x}_i)$ values to generate outcome, so that the proportions of 1's and -1's are approximately 50%.).

We divided the generated datasets into three folds and each time used two folds as training data and the other fold as testing data. The number of maximum iterations T is important to PKB, as using a large T will likely induce overfitting on training data and poor prediction on testing data. Therefore, we performed nested cross validation within the training data to select T. We further divided the training data into three folds and each time trained the PKB model using two folds while monitoring the loss function on the other fold at every iteration. Eventually, we identified the iteration number T^* with the minimum averaged loss on testing data and applied PKB to the whole training dataset up to T^* iterations.

We first evaluated the ability of PKB to correctly identify relevant pathways. For each simulation scenario, we calculated the average optimal weights across different cross validation runs and the results are shown in Figure 1, where the X-axis represents different pathways and the length of bars above them represents corresponding weights in the prediction functions. Note that for the underlying Model 1 and Model 2, only the first three pathways were relevant to the outcome, and in Model 3, the first ten pathways were relevant. In all the cases, PKB successfully assigned the largest weights to relevant pathways. Since PKB is an iterative approach, at some iterations, certain pathways irrelevant to the outcome may be selected by chance and added to the prediction function. This explains the non-zero weights of the irrelevant pathways and their values are clearly smaller than those of relevant pathways.

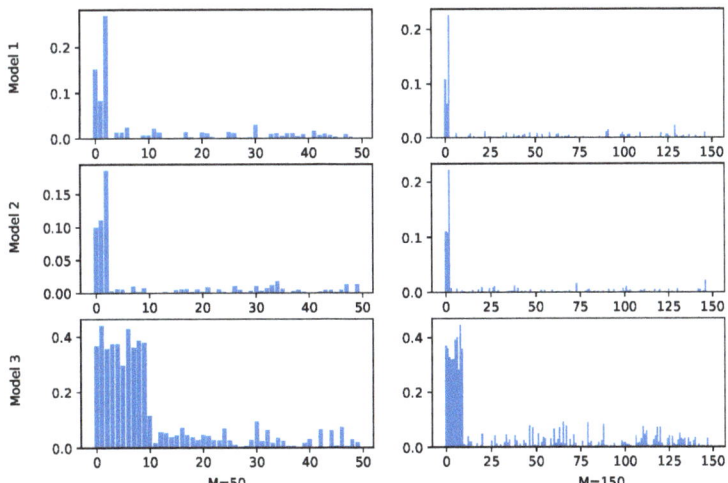

Figure 1. Estimated pathway weights by PKB in simulation studies. The X-axis represents pathways and the Y-axis represents estimated weights. Based on the simulation settings, the first three pathways are relevant in Models 1 and 2 and the first ten pathways are relevant in Model 3. M represents the number of simulated pathways.

We also applied several commonly used methods to the simulated datasets and compared their prediction accuracy with PKB. These methods included both non-pathway-based methods: Random Forest [25] and SVM [20] and pathway-based methods: NPR [11] and EasyMKL [14]. Model parameters we used for the above methods are listed in Section 3 of the Supplementary Materials. We used the same three-fold split of the data, as we used when applying PKB, to perform cross-validations for each competing method. The average prediction performance of the methods is summarized in Table 2. It can be seen that the pathway-based methods generally performed better than the non-pathway-based methods in all simulated scenarios. Among the pathway-based methods, the one that utilized kernels (EasyMKL) had comparable performances with the tree-based NPR method in Models 1 and 2 but had clearly superior performance in Model 3. This was likely due to the functional form of the log odds function $F(\mathbf{x})$ of Model 3. Note that genes in relevant pathways were involved in $F(\mathbf{x})$ in terms of their L_2 norms, which is hard to approximate by regression tree functions but can be well captured using kernel methods. In all scenarios, the best performance was achieved by one of the PKB methods. In four out of six scenarios, the PKB-L_2 method produced the smallest prediction errors, while in the other two scenarios, PKB-L_1 was slightly better. Although PKB-L_1 and PKB-L_2 had similar performances, PKB-L_1 was usually computationally faster, because in the optimization step of each iteration, the L_1 algorithm only looked for sparse solution of β's, which can be done more efficiently than PKB-L_2, which involves matrix inverse.

Table 2. Classification error rate from PKB and competing methods in simulation studies. The numbers below each model represent the number of pathways simulated in the data sets.

Method	Model 1		Model 2		Model 3	
	50	150	50	150	50	150
PKB-L_1	**0.151**	0.196	**0.198**	0.189	0.179	0.21
PKB-L_2	0.158	**0.185**	0.201	**0.183**	**0.157**	**0.173**
Random Forest	0.305	0.331	0.290	0.328	0.341	0.400
SVM	0.353	0.431	0.412	0.476	0.431	0.492
NPR	0.271	0.321	0.299	0.317	0.479	0.440
EasyMKL	0.253	0.284	0.268	0.330	0.212	0.300

3.2. Real Data Applications

We applied PKB to gene expression profiles to predict clinical features in three cancer studies, including breast cancer, melanoma and glioma. The clinical variables we considered included tumor grade, tumor site and metastasis status, which were all of great importance to cancer.

We used three commonly used pathway databases: KEGG, Biocarta and Gene Ontology (GO) Biological Process pathways. These databases provide lists of pathways with emphasis on different biological aspects, including molecular interactions and involvement in biological processes. The number of pathways from these databases ranges from 200 to 700. There is considerable overlap between pathways. To eliminate redundant information and control the overlap between pathways, we applied a preprocessing step to the databases with details provided in the Supplementary Materials Section 4.2.

Similar to the simulation studies, we compared the performances from different methods based on three fold cross validations following the same procedure as elaborated in Section 3.1. Most of the methods we considered have tuning parameters. We searched through different parameter configurations and reported the best result from cross-validation for each method. More details of the data sets and the implementations can be found in the Supplementary Materials Section 4. Table 3 shows the classification error rates from all methods. The numbers in bold are the optimal error rates for each column separately. In four out of five classifications, PKB was the best method (usually with the L_1 and L_2 methods being the top two). In the other case (melanoma, stage), NPR yielded the best results, with the PKB methods still ranking second and third.

We provide more detailed introductions to the data sets and clinical variables and interpretations of results by PKB in the following. For brevity of the article, we focus on presenting results for three outcomes, one from each data set and leave the other two in the Supplementary Materials (Section 4.4).

Table 3. Classification error rates on real data. The names in the parenthesis of each data set are the variables used as classification outcome. The best error rates are highlighted with bold font for each column.

Method	Data Sets				
	Metabric (Grade)	Glioma (Grade)	Glioma (Site)	Melanoma (Stage)	Melanoma (Met)
PKB-L_1	**0.274**	**0.283**	0.168	0.304	**0.081**
PKB-L_2	0.304	**0.283**	**0.154**	0.307	0.083
Random Forest	0.306	0.302	0.306	0.320	0.136
SVM	0.285	0.292	0.185	0.314	0.083
NPR	0.306	0.298	0.197	**0.282**	0.110
EasyMKL	0.297	0.302	0.291	0.314	0.100

3.2.1. Breast Cancer

Metabric is a breast cancer study that involved more than 2000 patients with primary breast tumors [26]. The data set provides copy number aberration, gene expression, mutation and long-term clinical follow-up information. We are interested in the clinical variable of tumor grade, which measures the abnormality of the tumor cells compared to normal cells under a microscope. It takes a value of 1, 2, or 3. Higher Grade indicates more abnormality and higher risk of rapid tumor proliferation. Since grade 1 contained the fewest samples, we pooled it together with Grade 2 as one class and treated Grade 3 as the other class.

We then applied PKB to samples in subtype Lum B, where the sample sizes for the two classes were most balanced (259 Grade 3 patients; 211 Grade 1,2 patients). For input gene expression data, we used the normalized mRNA expression (microarray) data for 24,368 genes provided in the data set. The model using GO Biological Process pathways and radial basis function (rbf) kernel yielded the best performance (error rate 27.4%). To obtain the pathways most relevant to tumor grade, we calculated the average pathway weights from the cross validation and sorted them from highest to lowest. Top fifteen pathways with the highest weights are presented in the first columns of Table 4.

Among all pathways, the cell aggregation and sequestering of metal ion pathways are the top two pathways in terms of the estimated pathway weights. Previous research has shown that cell aggregation contributes to the inhibition of cell death and anoikis-resistance, thereby promoting tumor cell proliferation. Genes in the cell aggregation pathway include TGFB2, MAPK14, FGF4 and FGF6, which play important roles in the regulation of cell differentiation and fate [27]. Moreover, the majority of genes in the sequestering of metal ion pathway encode calcium-binding proteins, which regulates calcium level and different cell signaling pathways relevant to tumorigenesis and progression [28]. Among these genes, S100A8 and S100A9 have been identified as novel diagnostic markers of human cancer [29]. The results suggest that PKB has identified pathways that are likely relevant to breast cancer grade.

3.2.2. Lower Grade Glioma

Glioma is a type of cancer developed in the glial cells in brain. As glioma tumor grows, it compresses normal brain tissue and can lead to disabling or fatal results. We applied our method to a lower Grade glioma data set from TCGA, where only grades 2 and 3 samples were collected (Grade 4 glioma, also known as glioblastoma, is studied in a separate TCGA study.) [30]. After removal of missing values, the numbers of patients in the cohort with grades 2 and 3 tumors were 248 and 265, respectively. We used Grade as the outcome variable to be classified and applied PKB with different parameter configurations. After cross validation, PKB using the third order polynomial (poly3) kernel and the GO Biological Process pathways yielded an error rate of 28.3%, which was the smallest among all methods. The top fifteen pathways selected in the model are listed in the second column of Table 4.

Table 4. Top fifteen pathways with the largest weights fitted by PKB. In each column, pathways are sorted in descending order from top to bottom. Pathways in the first two columns are from GO Biological Process pathways and the third column from Biocarta.

	Metabric (Grade)	Glioma (Grade)	Melanoma (Met)
1	Cell aggregation	Homophilic cell adhesion via plasma membrane adhesion molecules	Lectin induced complement pathway
2	Sequestering of metal ion	Neuropeptide signaling pathway	Classical complement pathway
3	Glutathione derivative metabolic process	Multicellular organismal macromolecule metabolic process	Phospholipase c delta in phospholipid associated cell signaling
4	Antigen processing and presentation of exogenous peptide antigen via mhc class i	Peripheral nervous system neuron differentiation	Fc epsilon receptor i signaling in mast cells
5	Sterol biosynthetic process	Positive regulation of hair cycle	Inhibition of matrix metalloproteinases
6	Pyrimidine containing compound salvage	Peptide hormone processing	Regulation of map kinase pathways through dual specificity phosphatases
7	Protein dephosphorylation	Hyaluronan metabolic process	Estrogen responsive protein efp controls cell cycle and breast tumors growth
8	Homophilic cell adhesion via plasma membrane adhesion molecules	Positive regulation of synapse maturation	Chaperones modulate interferon signaling pathway
9	Cyclooxygenase pathway	Stabilization of membrane potential	Il-10 anti-inflammatory signaling pathway
10	Establishment of protein localization to endoplasmic reticulum	Lymphocyte chemotaxis	Reversal of insulin resistance by leptin
11	Negative regulation of dephosphorylation	Insulin secretion	Bone remodeling
12	Xenophagy	Positive regulation of osteoblast proliferation	Cycling of ran in nucleocytoplasmic transport
13	Attachment of spindle microtubules to kinetochore	Negative regulation of dephosphorylation	Alternative complement pathway
14	Fatty acyl coa metabolic process	Trophoblast giant cell differentiation	Cell cycle: g^2/m checkpoint
15	Apical junction assembly	Synaptonemal complex organization	Hop pathway in cardiac development

The estimated pathway weights indicate that the cell adhesion pathway and the neuropeptide signaling pathway have the strongest association with glioma grade. Genes in the cell adhesion pathways generally govern the activities of cell adhesion molecules. Turning off the expression of cell-cell adhesion molecules is one of the hallmarks of tumor cells, by which tumor cells can inhibit antigrowth signals and promote proliferation. Previous studies have shown that deletion of carcinoembryonic antigen-related cell adhesion molecule 1 (CEACAM1) gene can contribute to cancer progression [31]. Cell Adhesion Molecule 1 (CADM1), CADM2, CADM3 and CADM4, serve as tumor suppressors and can inhibit cancer cell proliferation and induce apoptosis. Neuropeptide signaling pathway has also been implicated in tumor growth and progression. Neuropeptide Y is highly relevant to tumor cell proliferation and survival. Two NPY receptors, Y2R and Y5R, are also members of the neuropeptide signaling pathway. They are considered as important stimulatory mediators in tumor cell proliferation [32].

3.2.3. Melanoma

The next application of PKB is to a TCGA cutaneous melanoma dataset [33]. Melanoma is most often discovered after it has metastasized and the skin melanoma site is never found. Therefore, the majority of the samples are metastatic. In this data set, there are 369 metastatic samples and 103 primary samples. It is of great interest to study the genomic differences between the two types, thus we applied PKB to this data using metastatic/primary as the outcome variable. Using the Biocarta pathways and rbf kernel produced the smallest classification error rate (8.1%) among all methods. Fifteen pathways that PKB found most relevant to the outcome are presented in the third column of Table 4.

Two complement pathways, lectin induced complement pathway and classical complement pathway, came out from the PKB model as the most significant pathways. Proteins in complement system participate in a variety of biological processes of metastasis, such as epithelial-mesenchymal transition (EMT). EMT is an important process in the initiation stage of metastasis, through which cells in primary tumor lose cell-cell adhesion and gain invasive properties. Complement activation by tumor cells can recruit stromal cells to the tumor and induce EMT. Furthermore, complement proteins can mediate the degradation of extracellular matrix, thereby promoting tumor metastasis [34].

4. Discussion

In this paper, we have introduced the PKB model as a method to perform classification analysis of gene expression data, as well as identify pathways relevant to the clinical outcomes of interest. PKB usually yields sparse models in terms of the number of pathways, which enhances interpretability of the results. Moreover, the pathway weights as defined in Section 2 can be used as a measure of pathway importance and provides guidance for further experimental verifications.

Two types of regularizations are introduced in the optimization step of PKB, in order to select simple model with good fitting. Computation efficiency of the two methods depends on the regularization strengh: when regularization is strong, the L_1 method enjoys a computational advantage due to the sparsity of its solution; when regularization is weak, it requires more iterations to converge and yields worse run time than the L_2. In simulations and real data applications, both methods yielded comparable prediction accuracy. It is worth mentioning that the second-order approximation of the log loss function is also necessary for efficiency of PKB. The approximation yields an expression that is quadratic in terms of coefficients β, which allows the problem to be converted to LASSO or Ridge Regression after regularizations are added. If the original loss function was used, solving β would be more time consuming. In the applications, we only considered gene expression data as model input. However, our method can be easily generalized to use other continuous inputs, such as gene methylation measurements. By incorporating other properly designed kernel functions, it is also possible to handle discrete inputs (for example, the weighted IBS kernel for SNP data [7]).

There are several limitations of the current PKB approach. First of all, when constructing base learners from pathways, we use fixed bandwidth parameters (inverse of the number of genes in each pathway) in the kernel functions. Ideally, we would like the model to auto-determine the parameters. However, the number of such parameters is equal to the number of pathways, which is often too large to tune efficiently. Therefore, it remains a challenging task for future research. Second, we currently only use pathway as a criterion to group genes and within each pathway, all genes are treated equally. It is conceivable that the genes interact with each other through an underlying interaction network and intuitively, genes in the hub should get more weights compared to genes on the periphery. With the network information available, it is possible to build more sensible kernel functions as base learners [17]. Third, the pathway databases only cover a subset of the input genes. Both KEGG and Biocarta only include a few thousands of genes, while the number of input genes is usually beyond 15,000. Large number of genes, with the potential to provide additional prediction power, remain unused in the model. In our applications, we tried pooling together all unused genes and consider them as a new pathway but it did not significantly improve the results. Although genes annotated with pathways are supposed to be most informative, it is still worth looking for smarter ways of handling unannotated genes.

Supplementary Materials: Supplementary materials and reproduction code are available online at https://github.com/zengliX/PKB. Reproduction-related input data sets are available upon request from the corresponding author (hongyu.zhao@yale.edu).

Author Contributions: Conceptualization, L.Z. and H.Z.; methodology, L.Z.; software, L.Z. and Z.Y.; writing—original draft preparation, L.Z. and Z.Y.; writing—review and editing, all listed authors; supervision, H.Z.; project administration, L.Z. and H.Z.

Acknowledgments: This research was funded in part by NIH grant numbers P01 CA154295 and P50 CA196530.

Conflicts of Interest: The authors declare no conflict of interest.

Abbreviations

The following abbreviations are used in this manuscript:

KEGG   Kyoto Encyclopedia of Genes and Genomes
TCGA   The Cancer Genome Atlas

References

1. Subramanian, A.; Tamayo, P.; Mootha, V.K.; Mukherjee, S.; Ebert, B.L.; Gillette, M.A.; Paulovich, A.; Pomeroy, S.L.; Golub, T.R.; Lander, E.S.; et al. Gene set enrichment analysis: A knowledge-based approach for interpreting genome-wide expression profiles. *Proc. Natl. Acad. Sci. USA* **2005**, *102*, 15545–15550. [CrossRef] [PubMed]
2. Carlson, C.S.; Eberle, M.A.; Kruglyak, L.; Nickerson, D.A. Mapping complex disease loci in whole-genome association studies. *Nature* **2004**, *429*, 446. [CrossRef] [PubMed]
3. Kanehisa, M.; Goto, S. KEGG: Kyoto encyclopedia of genes and genomes. *Nucleic Acids Res.* **2000**, *28*, 27–30. [CrossRef] [PubMed]
4. Schaefer, C.F.; Anthony, K.; Krupa, S.; Buchoff, J.; Day, M.; Hannay, T.; Buetow, K.H. PID: The pathway interaction database. *Nucleic Acids Res.* **2008**, *37*, D674–D679. [CrossRef] [PubMed]
5. Nishimura, D. BioCarta. *Biotech Softw. Internet Rep. Comput. Softw. J. Sci.* **2001**, *2*, 117–120. [CrossRef]
6. Liu, D.; Lin, X.; Ghosh, D. Semiparametric Regression of Multidimensional Genetic Pathway Data: Least-Squares Kernel Machines and Linear Mixed Models. *Biometrics* **2007**, *63*, 1079–1088. [CrossRef]
7. Wu, M.C.; Lee, S.; Cai, T.; Li, Y.; Boehnke, M.; Lin, X. Rare-variant association testing for sequencing data with the sequence kernel association test. *Am. J. Hum. Genet.* **2011**, *89*, 82–93. [CrossRef]
8. Shou, J.; Massarweh, S.; Osborne, C.K.; Wakeling, A.E.; Ali, S.; Weiss, H.; Schiff, R. Mechanisms of tamoxifen resistance: Increased estrogen receptor-HER2/neu cross-talk in ER/HER2–positive breast cancer. *J. Natl. Cancer Inst.* **2004**, *96*, 926–935. [CrossRef]
9. Shtivelman, E.; Hensing, T.; Simon, G.R.; Dennis, P.A.; Otterson, G.A.; Bueno, R.; Salgia, R. Molecular pathways and therapeutic targets in lung cancer. *Oncotarget* **2014**, *5*, 1392. [CrossRef]
10. Berk, M. Neuroprogression: Pathways to progressive brain changes in bipolar disorder. *Int. J. Neuropsychopharmacol.* **2009**, *12*, 441–445. [CrossRef]
11. Wei, Z.; Li, H. Nonparametric pathway-based regression models for analysis of genomic data. *Biostatistics* **2007**, *8*, 265–284. [CrossRef] [PubMed]
12. Luan, Y.; Li, H. Group additive regression models for genomic data analysis. *Biostatistics* **2007**, *9*, 100–113. [CrossRef] [PubMed]

13. Gönen, M.; Margolin, A.A. Drug susceptibility prediction against a panel of drugs using kernelized Bayesian multitask learning. *Bioinformatics* **2014**, *30*, i556–i563. [CrossRef] [PubMed]
14. Aiolli, F.; Donini, M. EasyMKL: A scalable multiple kernel learning algorithm. *Neurocomputing* **2015**, *169*, 215–224. [CrossRef]
15. Costello, J.C.; Heiser, L.M.; Georgii, E.; Gönen, M.; Menden, M.P.; Wang, N.J.; Bansal, M.; Hintsanen, P.; Khan, S.A.; Mpindi, J.P.; et al. A community effort to assess and improve drug sensitivity prediction algorithms. *Nat. Biotechnol.* **2014**, *32*, 1202–1212. [CrossRef] [PubMed]
16. Friedrichs, S.; Manitz, J.; Burger, P.; Amos, C.I.; Risch, A.; Chang-Claude, J.; Wichmann, H.E.; Kneib, T.; Bickeböller, H.; Hofner, B. Pathway-based kernel boosting for the analysis of genome-wide association studies. *Comput. Math. Methods Med.* **2017**, *2017*. [CrossRef] [PubMed]
17. Manica, M.; Cadow, J.; Mathis, R.; Martínez, M.R. PIMKL: Pathway-Induced Multiple Kernel Learning. *NPJ Syst. Biol. Appl.* **2019**, *5*, 8. [CrossRef] [PubMed]
18. Friedman, J.H. Greedy function approximation: A gradient boosting machine. *Ann. Stat.* **2001**, *29*, 1189–1232. [CrossRef]
19. Friedman, J.; Hastie, T.; Tibshirani, R. Additive logistic regression: A statistical view of boosting (with discussion and a rejoinder by the authors). *Ann. Stat.* **2000**, *28*, 337–407. [CrossRef]
20. Guyon, I.; Weston, J.; Barnhill, S.; Vapnik, V. Gene selection for cancer classification using support vector machines. *Mach. Learn.* **2002**, *46*, 389–422. [CrossRef]
21. Drucker, H.; Burges, C.J.; Kaufman, L.; Smola, A.J.; Vapnik, V. Support Vector Regression Machines. In *Advances in Neural Information Processing Systems*; MIT Press Cambridge, MA, USA, 1997; pp. 155–161.
22. Fukumizu, K.; Bach, F.R.; Jordan, M.I. Kernel dimension reduction in regression. *Ann. Stat.* **2009**, *37*, 1871–1905. [CrossRef]
23. Friedman, J.; Hastie, T.; Tibshirani, R. *The Elements of Statistical Learning*; Springer Series in Statistics: New York, NY, USA, 2001; Volume 1.
24. Johnson, R.; Zhang, T. Learning nonlinear functions using regularized greedy forest. *IEEE Trans. Pattern Anal. Mach. Intell.* **2014**, *36*, 942–954. [CrossRef] [PubMed]
25. Breiman, L. Random forests. *Mach. Learn.* **2001**, *45*, 5–32. [CrossRef]
26. Pereira, B.; Chin, S.F.; Rueda, O.M.; Vollan, H.K.M.; Provenzano, E.; Bardwell, H.A.; Pugh, M.; Jones, L.; Russell, R.; Sammut, S.J.; et al. The somatic mutation profiles of 2,433 breast cancers refines their genomic and transcriptomic landscapes. *Nat. Commun.* **2016**, *7*, 11479. [CrossRef] [PubMed]
27. Zhang, X.; Xu, L.h.; Yu, Q. Cell aggregation induces phosphorylation of PECAM-1 and Pyk2 and promotes tumor cell anchorage-independent growth. *Mol. Cancer* **2010**, *9*, 7. [CrossRef] [PubMed]
28. Monteith, G.R.; McAndrew, D.; Faddy, H.M.; Roberts-Thomson, S.J. Calcium and cancer: Targeting Ca^{2+} transport. *Nat. Rev. Cancer* **2007**, *7*, 519. [CrossRef]
29. Hermani, A.; Hess, J.; De Servi, B.; Medunjanin, S.; Grobholz, R.; Trojan, L.; Angel, P.; Mayer, D. Calcium-binding proteins S100A8 and S100A9 as novel diagnostic markers in human prostate cancer. *Clin. Cancer Res.* **2005**, *11*, 5146–5152. [CrossRef]
30. TCGA. Comprehensive, integrative genomic analysis of diffuse lower-grade gliomas. *N. Engl. J. Med.* **2015**, *2015*, 2481–2498.
31. Leung, N.; Turbide, C.; Olson, M.; Marcus, V.; Jothy, S.; Beauchemin, N. Deletion of the carcinoembryonic antigen-related cell adhesion molecule 1 (Ceacam1) gene contributes to colon tumor progression in a murine model of carcinogenesis. *Oncogene* **2006**, *25*, 5527. [CrossRef]
32. Tilan, J.; Kitlinska, J. Neuropeptide Y (NPY) in tumor growth and progression: Lessons learned from pediatric oncology. *Neuropeptides* **2016**, *55*, 55–66. [CrossRef]
33. TCGA; Akbani, R.; Akdemir, K.C.; Aksoy, B.A.; Albert, M.; Ally, A.; Amin, S.B.; Arachchi, H.; Arora, A.; Auman, J.T.; et al. Genomic classification of cutaneous melanoma. *Cell* **2015**, *161*, 1681–1696. [CrossRef] [PubMed]
34. Pio, R.; Corrales, L.; Lambris, J.D. The Role of Complement in Tumor Growth. In *Tumor Microenvironment and Cellular Stress*; Springer: Berlin, Germany, 2014; pp. 229–262.

© 2019 by the authors. Licensee MDPI, Basel, Switzerland. This article is an open access article distributed under the terms and conditions of the Creative Commons Attribution (CC BY) license (http://creativecommons.org/licenses/by/4.0/).

Article

Testing Differential Gene Networks under Nonparanormal Graphical Models with False Discovery Rate Control

Qingyang Zhang

Department of Mathematical Sciences, University of Arkansas, Arkansas, AR 72701, USA; qz008@uark.edu

Received: 20 January 2020; Accepted: 30 January 2020; Published: 5 February 2020

Abstract: The nonparanormal graphical model has emerged as an important tool for modeling dependency structure between variables because it is flexible to non-Gaussian data while maintaining the good interpretability and computational convenience of Gaussian graphical models. In this paper, we consider the problem of detecting differential substructure between two nonparanormal graphical models with false discovery rate control. We construct a new statistic based on a truncated estimator of the unknown transformation functions, together with a bias-corrected sample covariance. Furthermore, we show that the new test statistic converges to the same distribution as its oracle counterpart does. Both synthetic data and real cancer genomic data are used to illustrate the promise of the new method. Our proposed testing framework is simple and scalable, facilitating its applications to large-scale data. The computational pipeline has been implemented in the R package *DNetFinder*, which is freely available through the Comprehensive R Archive Network.

Keywords: gene regulatory network; nonparanormal graphical model; network substructure; false discovery rate control

---

1. Background

Inferring the structural change of a network under different conditions is essential in many problems arising in biology, medicine, and other scientific fields. For instance, in genomics, it is often of importance to study the structural change of a genetic pathway between diseased and normal groups. In the field of brain mapping, it is critical to identify the difference in brain connectivity between groups (for example, the brain connectivity network of normal subjects and patients often possess different structures). Most of these applications have relied on the prevailing Gaussian graphical models (GGMs) because of its good interpretability and computational convenience, and there is a rich and growing literature on learning differential networks under GGMs. To name a few, Guo et al. (2015) [1] introduced a joint estimation for multiple GGMs by a group lasso approach, under the assumption that the GGMs being studied are sparse and only differ in a small portion of edges. Danaher et al. (2014) [2] proposed a fused graphical lasso method which is free from the sparsity assumption on condition-specific networks and only requires the sparsity of the differential network. Zhao et al. (2014) [3] constructed a new estimator which directly estimates the differential network defined as $\Delta = \Sigma_X^{-1} - \Sigma_Y^{-1}$, where Σ_X^{-1} and Σ_Y^{-1} represent the two condition-specific precision matrices and $\Delta, \Sigma_X^{-1}, \Sigma_Y^{-1}$ have the same dimension. Liu (2017) [4] presented a new test to simultaneously study structural similarities and differences between multiple high-dimensional GGMs, which adopts the partial correlation coefficients to characterize the potential changes of dependency strength between two variables.

Most of the aforementioned algorithms were based upon penalized likelihood maximization. Although some algorithms were consistent under certain regularity conditions, they failed to control the false discovery rate (FDR) of the substructure detection as it is difficult to choose a tuning parameter to control the FDR at the desired level [1–3]. One exception is Liu (2017), who introduced a hierarchical testing framework to adjust for the multiplicity. Liu's test was constructed to asymptotically control the FDR while keeping satisfactory statistical power. Simulation studies in [4] have shown that this new test exhibits substantial power gains over existing methods such as graphical lasso. One major drawback that limits the application of Liu's test is the Gaussian assumption, which is often violated in practice especially in genomics. For instance, some digital measurements of gene expression level such as RNA-Seq data often greatly deviate from normality even after log-transformation or other variance-stabilizing transformations. In this paper, we aim to extend Liu's work to a more flexible semiparametric framework, namely the nonparanormal graphical models (NPNGMs), where the random variables are assumed to follow a multivariate normal distribution after a set of monotonically increasing transformations. We use a novel rank-based multiple testing method to detect the structural difference between multiple networks from non-Gaussian data. The method is computationally efficient and asymptotically controls the FDR at a desired level. To begin with, we give the formal definition of nonparanormal distribution:

Definition 1. *A random vector $Y = (Y_1, Y_2, ..., Y_p)$ follows a nonparanormal distribution if there exists a set of univariate and monotonically increasing transformations, $f = (f_1, ..., f_p)$, such that:*

$$(X_1, ..., X_p) \equiv (f_1(Y_1), ..., f_p(Y_p)) \sim N(\boldsymbol{\mu}, \boldsymbol{\Sigma}),$$

where $\boldsymbol{\mu}$ and $\boldsymbol{\Sigma}$ denote the mean and covariance matrix in the multivariate normal distribution, respectively. The distribution of Y depends on three parameters and it can be generally written as $Y \sim NPN(\boldsymbol{\mu}, \boldsymbol{\Sigma}, f)$.

By Definition 1 and Sklar's theorem, it is easy to verify that when the transformation functions f_j's are all differentiable, the nonparanormal distribution $NPN(\boldsymbol{\mu}, \boldsymbol{\Sigma}, f)$ is equivalent to a Gaussian copula [5]. As graphical models, the NPNGMs are much more flexible than GGMs in modeling non-Gaussian data while retaining the interpretability of the latter. Some recent studies have established the estimation and properties of high dimensional nonparanormal graphical models. For example, Liu et al. (2009) [5], who first studied high-dimensional NPNGMs, bridged the estimations of GGMs and NPNGMs by a nonparametric and truncated (Winsorized) estimator of the unknown transformation functions. Xue and Zou (2012) [6] proposed to use an adjusted Spearman's correlation to estimate the structure of high-dimensional NPNGMs, and they showed that the rank-based estimator achieves the same rate of convergence as its oracle counterpart (i.e., assuming known transformation functions). Despite the advances in single NPNGM estimation, to the best of our knowledge, the inference of differential substructure between multiple NPNGMs has not been studied. In this paper, we tackled this problem by embedding the Winsorized estimator into the testing framework of Liu (2017). Under some regularity conditions, we showed that the new test statistic converges to the same distribution as its oracle counterpart does [4].

We begin with the notations and problem formulation. For a vector $a = (a_1, ..., a_p)$, we define its ℓ_0 norm as $\|a\|_{\ell_0} = \sum_{i=1}^p I\{a_i \neq 0\}$, its ℓ_1 norm as $\|a\|_{\ell_1} = \sum_{i=1}^p |a_i|$, its ℓ_2 norm as $\|a\|_{\ell_2} = \sqrt{\sum_{i=1}^p a_i^2}$, and its ℓ_∞ norm as $\|a\|_{\ell_\infty} = \max_i |a_i|$. For a matrix $A = (a_{ij}) \in \mathbb{R}^{p \times q}$, we define its ℓ_0 norm as $\|A\|_0 = \sum_{i,j} I\{a_{ij} \neq 0\}$, its ℓ_1 norm as $\|A\|_1 = \sum_{i,j} |a_{ij}|$, its Frobenius norm as $\|A\|_F = \sqrt{\sum_{i,j} a_{ij}^2}$, and its ℓ_∞ norm as $\|A\|_\infty = \max_{i,j} |a_{ij}|$. Let $A_{i,-j}$ denote the ith row of A with its jth entry being removed and $A_{-i,j}$ denote the jth column with its ith entry being removed. We use $A_{-i,-j}$ to denote a $(p-1) \times (q-1)$ matrix by removing the ith row and the jth column. For square matrix B, we let $\lambda_{\max}(B)$ and $\lambda_{\min}(B)$ denote the largest and smallest eigenvalues of B respectively. In addition, for a given sequence of random variable $\{X_n, n = 1, 2, ...\}$ and a constant sequence $\{a_n, n = 1, 2, ...\}$, $X_n = o_p(a_n)$ denotes that X_n/a_n converges to zero in probability as n approaches to infinity and $X_n = O_p(a_n)$ denotes that X_n/a_n is stochastically bounded. If there are positive constants c and C such that $c \leq X_n/a_n \leq C$ for all $n \geq 1$, we write $X_n \sim a_n$.

To formulate the problem, we let $k \in \{1, 2, ..., K\}$ be the index of class, p be the dimension, and $(Y_1^{(k)}, ..., Y_{n_k}^{(k)})$ be a sample of size n_k for class k where $Y_m^{(k)} = (Y_{m1}^{(k)}, ..., Y_{mp}^{(k)})^T \in \mathbb{R}^p$, $m \in \{1, ..., n_k\}$. Under $Y_m^{(k)} \sim NPN(\boldsymbol{\mu}^{(k)}, \boldsymbol{\Sigma}^{(k)}, \boldsymbol{f}^{(k)})$, we test the following hypothesis:

$$H_{0ij} : \rho_{ij.}^{(1)} = \rho_{ij.}^{(2)} = ... = \rho_{ij.}^{(K)},$$
$$H_{aij} : \rho_{ij.}^{(k)} \neq \rho_{ij.}^{(k')}, \text{ for some } k, k' \in \{1, ..., K\},$$

where $1 \leq i, j \leq p$, $\{\boldsymbol{\Sigma}^{(k)}\}^{-1} = \boldsymbol{\Omega}^{(k)} = (\omega_{ij}^{(k)})$, and $\rho_{ij.}^{(k)}$ represents the partial correlation coefficient between $X_i^{(k)}$ and $X_j^{(k)}$ given $\mathbf{X}^{(k)} \setminus (X_i^{(k)}, X_j^{(k)})$, $(X_{m1}^{(k)}, ..., X_{mp}^{(k)}) = (f_1^{(k)}(Y_{m1}^{(k)}), ..., f_p^{(k)}(Y_{mp}^{(k)}))$. The edge (i, j) is a differential edge if $\rho_{ij.}^{(k)} \neq \rho_{ij.}^{(k')}$ for some $k, k' \in \{1, ..., K\}$, and the differential network is defined as the set of all differential edges. As a well-known result in statistics, $\rho_{ij.}^{(k)} = -\omega_{ij}^{(k)} / \sqrt{\omega_{ii}^{(k)} \omega_{jj}^{(k)}}$. Here, we consider an equivalent alternative of the hypothesis testing above. Similar as in [4], let

$$S_{ij}(\boldsymbol{\Omega}) = \sqrt{\sum_{1 \leq k < k' \leq K} (\rho_{ij.}^{(k)} - \rho_{ij.}^{(k')})^2}, \quad (1)$$

then the hypothesis testing can be simplified as

$$H_{0ij} : S_{ij}(\boldsymbol{\Omega}) = 0,$$
$$H_{aij} : S_{ij}(\boldsymbol{\Omega}) > 0.$$

As $S_{ij}(\boldsymbol{\Omega}) = S_{ji}(\boldsymbol{\Omega})$, we define $\mathcal{H}_0 = \{H_{0ij}, 1 \leq i < j \leq p\}$ and $\mathcal{H}_a = \{H_{aij}, 1 \leq i < j \leq p\}$, and the total numbers of tests are $p(p-1)/2$, i.e., $card(\mathcal{H}_0) = card(\mathcal{H}_a) = p(p-1)/2$.

The rest of this paper is structured as follows: In Section 2, we introduce the new test statistic and multiple testing procedure. In Section 3 we perform a simulation study to evaluate the finite sample performance of the proposed test in terms of FDR control and statistical power. We then apply the new method to a rich genomic data to study the genetic difference between four breast cancer subtypes. We discuss the strength and shortcomings of the test in Section 5. Technical proof of the asymptotic results is provided in Appendix A.

2. Statistical Methods

2.1. Winsorized Estimator of the Latent Gaussian Variables

In practice, the transformation functions $\boldsymbol{f}^{(k)} = (f_1^{(k)}, ..., f_p^{(k)})$ in the nonparanormal distribution are unknown. However, one can use a Winsorized estimator to approximate $\boldsymbol{f}^{(k)}$, i.e., to impute the latent Gaussian variables (oracle data) $(X_{m1}^{(k)}, ..., X_{mp}^{(k)})_{1 \leq m \leq n_k}$. To illustrate the Winsorized estimator, we define the following quantile function:

$$\hat{h}_j^{(k)}(t) = \Phi^{-1}(\tilde{F}_j^{(k)}(t)), \quad 1 \leq j \leq p,$$

where $\tilde{F}_j^{(k)}$ is some estimator of the cumulative distribution function of $Y_j^{(k)}$, and a natural choice for $\tilde{F}_j^{(k)}$ would be the empirical cumulative distribution function (eCDF)

$$\hat{F}_j^{(k)}(t) = \frac{1}{n_k} \sum_{m=1}^{n_k} I\{Y_{mj}^{(k)} \leq t\}.$$

One major drawback of the eCDF above is that under high dimensionality, the variance of $\hat{F}_j^{(k)}(t)$ could be too large. To overcome the problem, Liu et al. (2009) considered a truncated (Winsorized) estimator as follows:

$$\tilde{F}_j^{(k)} = \begin{cases} \delta_n & \hat{F}_j^{(k)}(t) < \delta_n \\ \hat{F}_j^{(k)}(t) & \delta_n \le \hat{F}_j^{(k)}(t) \le 1 - \delta_n, \\ 1 - \delta_n & \hat{F}_j^{(k)}(t) > 1 - \delta_n \end{cases}$$

where δ_n serves as the truncation parameter that should be carefully chosen. Liu et al. (2009) [5] suggested $\delta_n = 1/(4n^{1/4}\sqrt{\pi \log n})$ to balance the bias and variance of eCDF, and so we will use this value in our calculations. To estimate the transformation functions and impute the latent Gaussian variable **X**, we define

$$X_{mj}^{(k)*} = \tilde{f}_j^{(k)}(Y_{mj}^{(k)}) = \hat{\mu}_j^{(k)} + \hat{\sigma}_j^{(k)} \tilde{h}_j^{(k)}(Y_{mj}^{(k)}),$$

where $\tilde{h}_j^{(k)}(t)$, $\hat{\mu}_j^{(k)}$ and $\hat{\sigma}_j^{(k)}$ are given below:

$$\tilde{h}_j^{(k)}(t) = \Phi^{-1}(\tilde{F}_j^{(k)}(t)),$$

$$\hat{\mu}_j^{(k)} = \frac{1}{n_k} \sum_{m=1}^{n_k} Y_{mj}^{(k)},$$

$$\hat{\sigma}_j^{(k)} = \sqrt{\frac{1}{n_k} \sum_{m=1}^{n_k} (Y_{mj}^{(k)} - \hat{\mu}_j^{(k)})^2}.$$

The Winsorized estimator $X_{mj}^{(k)*}$ generally works well in approximating the unknown $X_{mj}^{(k)}$, and it could be used to estimate the oracle sample covariance. Let $\hat{\Sigma}^{(k)}$ be the sample covariance matrix by the oracle data, and $\tilde{\Sigma}^{(k)}$ be the sample covariance matrix by $(X_1^{(k)*}, ..., X_p^{(k)*})$, that is

$$\tilde{\Sigma}^{(k)} = \frac{1}{n_k} \sum_{m=1}^{n_k} (X_m^{(k)*} - \tilde{\boldsymbol{\mu}}^{(k)})(X_m^{(k)*} - \tilde{\boldsymbol{\mu}}^{(k)})^T,$$

where $\tilde{\boldsymbol{\mu}}^{(k)} = (1/n_k)\sum_{m=1}^{n_k} X_m^{(k)*}$. Liu et al. (2009) established the following consistency results under mild regularity conditions:

$$\|\tilde{\Sigma}^{(k)} - \hat{\Sigma}^{(k)}\|_\infty = O_p\left(\sqrt{\frac{\log p \log^2 n_k}{n_k^{1/2}}}\right).$$

When estimating the precision matrix $\Omega^{(k)}$, one can consider a modified graphical lasso based on imputed data, i.e.,

$$\tilde{\Omega}_{glasso}^{(k)} = \arg\min_{\Omega} \left\{ \text{tr}(\Omega \tilde{\Sigma}^{(k)}) - \log|\Omega| + \lambda \|\Omega\|_1 \right\}. \qquad (2)$$

Liu et al. (2009) showed the following convergence, which elucidated the asymptotic equivalence between the oracle data and imputed data in the structural estimation of NPNGM

$$\|\tilde{\Omega}_{glasso}^{(k)} - \Omega^{(k)}\|_F = O_p\left(\sqrt{\frac{(\|\Omega^{(k)}\|_0 + p)\log p \log^2 n_k}{n_k^{1/2}}}\right).$$

2.2. Asymptotic Results for a Single Class

To extend Liu's test to a nonparanormal case, we first consider the problem of single GGM estimation based on oracle data, i.e., $(X_{m1}^{(k)}, ..., X_{mp}^{(k)})_{1 \le m \le n_k} \sim N(\mu_k, \Sigma_k)$, in the following regression framework

$$X_{mj}^{(k)} = \alpha_j^{(k)} + \mathbf{X}_{m,-j}^{(k)'} \boldsymbol{\beta}_j^{(k)} + \epsilon_{mj}^{(k)}. \tag{3}$$

It is not hard to show that the regression coefficients $\boldsymbol{\beta}_j^{(k)} = (\beta_{j,1}^{(k)}, ..., \beta_{j,j-1}^{(k)}, \beta_{j,j+1}^{(k)}, \beta_{j,p}^{(k)})$ and the error term $\epsilon_{mj}^{(k)}$ satisfy

$$\boldsymbol{\beta}_j^{(k)} = -(\omega_{jj}^{(k)})^{-1} \boldsymbol{\Omega}_{-j,j}^{(k)}, \quad \text{cov}(\epsilon_{mi}^{(k)}, \epsilon_{mj}^{(k)}) = \frac{\omega_{ij}^{(k)}}{\omega_{ii}^{(k)} \omega_{jj}^{(k)}}.$$

As the oracle data $(X_{m1}^{(k)}, ..., X_{mp}^{(k)})_{1 \le m \le n_k}$ in Equation (3) are generally unknown, we consider a new regression model based on Winsorized imputations:

$$X_{mj}^{(k)*} = \hat{\alpha}_j^{(k)} + \mathbf{X}_{m,-j}^{(k)*'} \hat{\boldsymbol{\beta}}_j^{(k)} + \epsilon_{mj}^{(k)*}. \tag{4}$$

In solving the problem of single GGM estimation, Liu (2017) proposed an elegant test based on a bias-corrected sample covariance. This has motivated us to construct the following new statistic

$$S_{ij}^{(k)*} = \sqrt{\frac{1}{n_k r_{ii}^{(k)*} r_{jj}^{(k)*}}} \left(\sum_{m=1}^{n_k} \epsilon_{mi}^{(k)*} \epsilon_{mj}^{(k)*} + \sum_{m=1}^{n_k} \{\epsilon_{mi}^{(k)*}\}^2 \hat{\beta}_{i,j}^{(k)} + \sum_{m=1}^{n_k} \{\epsilon_{mj}^{(k)*}\}^2 \hat{\beta}_{j,i}^{(k)} \right), \tag{5}$$

where $r_{ij}^{(k)*} = (1/n_k) \sum_{m=1}^{n_k} \epsilon_{mi}^{(k)*} \epsilon_{mj}^{(k)*}$. By letting $\bar{\boldsymbol{\epsilon}}^{(k)} = (1/n_k) \sum_{m=1}^{n_k} \boldsymbol{\epsilon}_m^{(k)}$, $(\hat{\sigma}_{ij,\epsilon}^{(k)})_{1 \le i,j \le p} = (1/n_k) \sum_{m=1}^{n_k} (\boldsymbol{\epsilon}_m^{(k)} - \bar{\boldsymbol{\epsilon}}^{(k)}) (\boldsymbol{\epsilon}_m^{(k)} - \bar{\boldsymbol{\epsilon}}^{(k)})'$, $b_{ij}^{(k)} = \omega_{ii}^{(k)} \hat{\sigma}_{ii,\epsilon}^{(k)} + \omega_{jj}^{(k)} \hat{\sigma}_{jj,\epsilon}^{(k)} - 1$, we will prove that, under mild conditions (see a detailed proof in Appendix A)

$$S_{ij}^{(k)*} + b_{ij}^{(k)} \frac{\omega_{ij}^{(k)}}{\omega_{ii}^{(k)} \omega_{jj}^{(k)}} \xrightarrow{D} N\left(0, 1 + \frac{\{\omega_{ij}^{(k)}\}^2}{\omega_{ii}^{(k)} \omega_{jj}^{(k)}}\right). \tag{6}$$

Similar as in [4], the estimated coefficients $\hat{\boldsymbol{\beta}}_j^{(k)}$ must satisfy the following conditions:

$$\|\hat{\boldsymbol{\beta}}_j^{(k)} - \boldsymbol{\beta}_j^{(k)}\|_{\ell_1} = O_p(a_n^{(k)}),$$

$$\min\left\{ \lambda_{\max}^{1/2}(\boldsymbol{\Sigma}^{(k)}) \|\hat{\boldsymbol{\beta}}_j^{(k)} - \boldsymbol{\beta}_j^{(k)}\|_{\ell_2}, \max_{1 \le j \le p} \sqrt{(\hat{\boldsymbol{\beta}}_j^{(k)} - \boldsymbol{\beta}_j^{(k)})^T \hat{\boldsymbol{\Sigma}}_{-j,-j}^{(k)} (\hat{\boldsymbol{\beta}}_j^{(k)} - \boldsymbol{\beta}_j^{(k)})} \right\} = O_p(b_n^{(k)}),$$

where

$$a_n^{(k)} = o(\sqrt{\log p / n_k}), \text{ and } b_n^{(k)} = o(n_k^{-1/4}). \tag{7}$$

Equation (6) is our main result, which is essentially a counterpart of Proposition 3.1 in [4]. The detailed proof is given in Appendix A. The asymptotic result we obtained here suggested that, by an appropriate choice of regression coefficients $\hat{\boldsymbol{\beta}}_j^{(k)}$, Liu's test can be readily extended to a nonparanormal framework by Winsorized imputation. Under GGMs, the condition (7) can be satisfied by several popular shrinkage estimators including lasso estimator and Dantzig selector. For the choice of $\boldsymbol{\beta}_j^{(k)}$ under NPNGMs, one can use the rank-based method introduced by Xue and Zou (2012) [6]. Xue and Zou (2012) showed that the rank-based estimator (e.g., rank-based lasso and rank-based Dantzig selector) achieved exactly the same convergence rate as its oracle counterpart, therefore, it also satisfies our condition (7).

2.3. Multiple Testing Procedure for FDR Control

Now we introduce the multiple testing procedure for FDR control based on the single-class result from Equation (6). As suggested in [4], the partial correlation coefficient can be well estimated by a thresholding estimator

$$\hat{\rho}_{ij\cdot}^{(k)} = S_{ij}^{(k)} I\{|S_{ij}^{(k)}| \geq 2\sqrt{\frac{\log p}{n_k}}\},$$

and we define the following two-sample test statistics

$$S_{ij}^{(k,k')} = \frac{S_{ij}^{(k)} - S_{ij}^{(k')}}{\sqrt{\frac{1}{n_k}(1 - \{\hat{\rho}_{ij\cdot}^{(k)}\}^2)^2 + \frac{1}{n_{k'}}(1 - \{\hat{\rho}_{ij\cdot}^{(k')}\}^2)^2}}.$$

In the multi-sample case $\boldsymbol{S}_{ij} = (S_{ij}^{(k,k')})_{1 \leq k < k' \leq K}$, we consider a sum squared test statistics

$$\mathbb{S}_{ij} = \sqrt{\sum_{k<k'}\{S_{ij}^{(k,k')}\}^2}.$$

Motivated by [4] (Equations 2.6 and 2.7) and [7], we define the following statistic

$$T_{ij} = \Phi^{-1} \mathrm{P}\left(\sqrt{\sum_{i=1}^{M} \lambda_i Z_i^2} \leq \mathbb{S}_{ij}\right),$$

and constant $A = (P_0 - \hat{P}_0)/Q_0$, where $Z_i, i = 1, \ldots, M$ represent a sequence of M i.i.d. standard normal random variables, $P_0 = 2\Phi(1) - 1$, $\hat{P}_0 = 2\sum_{1 \leq i < j \leq p} I\{|T_{ij}| \leq 1\}/(p^2 - p)$, $Q_0 = \sqrt{2}\phi(1)$ and $A(t) = (1 + |A|\frac{|t|\phi(t)}{\sqrt{2}(1-\Phi(t))})^{-1}$. For a given $0 < \alpha_0 < 1$, let

$$t(\alpha_0) = \inf\left\{t \in \mathbb{R}, 1 - \phi(t) \leq \frac{\alpha_0 A(t) \max\{1, \sum_{1 \leq i < j \leq p} I\{T_{ij} \geq t\}\}}{(p^2 - p)/2}\right\}.$$

the FDR can be controlled at level α, if we reject $H_{0ij} : S_{ij}(\boldsymbol{\Omega}) = 0$ when $T_{ij} \geq t(\alpha_0)$. One may refer to [7] for the detailed proof about this testing procedure.

Our proposed computational pipeline consisted of three steps: (1) Winsorized imputation for the latent Gaussian variables; (2) rank-based estimation of regression coefficients, and (3) multiple testing with FDR control. On the whole, we put forward a simple procedure to estimate the structural difference between multiple nonparanormal graphical models. The computational pipeline for a two-sample comparison has been implemented in the R package *DNetFinder*, which can be downloaded from the Comprehensive R Archive Network (CRAN).

3. Numerical Study

We performed a simulation study to evaluate the finite sample performance of the proposed procedure. In particular, we evaluated the empirical false discovery rate (eFDR) as well as the statistical power under two classes, i.e., $K = 2$. The dimension and sample size were set to be $p = 200$ and $n_1 = n_2 = 100$. We consider two commonly used graph-generating models including the band graph and Erdős–Rényi (ER) graph, and two estimators for regression coefficients including lasso estimator and Dantzig selector. Detailed set-up for precision matrices $\boldsymbol{\Omega}_1$ and $\boldsymbol{\Omega}_2$ are given below:

- **Band graph:** $\mathbf{\Omega}_1 = (\omega_{ij})_{1 \leq i,j \leq p}$ was obtained by the following assignments

$$\omega_{ij} = \begin{cases} 1 & |i-j| = 0 \\ 0.6 & |i-j| = 1 \\ 0 & |i-j| \geq 2 \end{cases}.$$

We then randomly picked 50 edges in $\mathbf{\Omega}_1$ as the differential edges and changed their signs in $\mathbf{\Omega}_2$. To ensure positive definiteness, we added $\max(|\lambda_{\min}(\mathbf{\Omega}_1)|, |\lambda_{\min}(\mathbf{\Omega}_2)|) + 0.05$, to the diagonal of $\mathbf{\Omega}_1$ and $\mathbf{\Omega}_2$.

- **Erdős–Rényi (ER) graph:** Each node pair (i, j) were randomly connected with probability 5%. A correlation coefficient is generated for each edge in the network from a two-part uniform distribution $[-1/2, -1/4] \cup [1/4, 1/2]$. To ensure positive-definiteness, we shrunk the correlations by a factor of 5 and the diagonals were set to be one for $\mathbf{\Omega}_1$. We then randomly selected 5% of the edges as the differential edges, and changed their signs in $\mathbf{\Omega}_2$.

For each graph, we generated the latent Gaussian data (oracle data) from $N(\mathbf{0}, \mathbf{\Omega}^{-1})$, $\mathbf{\Omega} \in \{\mathbf{\Omega}_1, \mathbf{\Omega}_2\}$, and a Winsorized estimator with truncation parameter $\delta_n = 1/(4n^{1/4}\sqrt{\pi \log n})$ was used to implement our test. The performance of the proposed method was then evaluated in two aspects: false discovery rate control and statistical power. In particular, we compared the results based on oracle data and imputed data by the Winsorized estimator. Two estimators including the lasso estimator and Dantzig selector were used to estimate coefficients $\hat{\boldsymbol{\beta}}$. For oracle data, we applied the R package *flare* to calculate the solution path over a sequence of 20 candidate λ's and tune by Akaike information criterion (AIC). For imputed data, we adopted the rank-based methods introduced by [6], i.e., the rank-based lasso and rank-based Dantzig selector. The simulation was repeated for 100 times for each FDR level ($\alpha \in \{0.05, 0.10, 0.15, ..., 0.50\}$) and the average empirical FDR and statistical power were summarized.

Figures 1 and 2 compared the empirical false discovery rate (eFDR) with the desired level α under the band graph and ER graph. It can be seen that the empirical FDR based on imputed data is close to the one by oracle data, both close to the desired level of α, suggesting that the FDRs were controlled quite well for both cases. The lasso estimator works almost equally well as Dantzig selector in both settings.

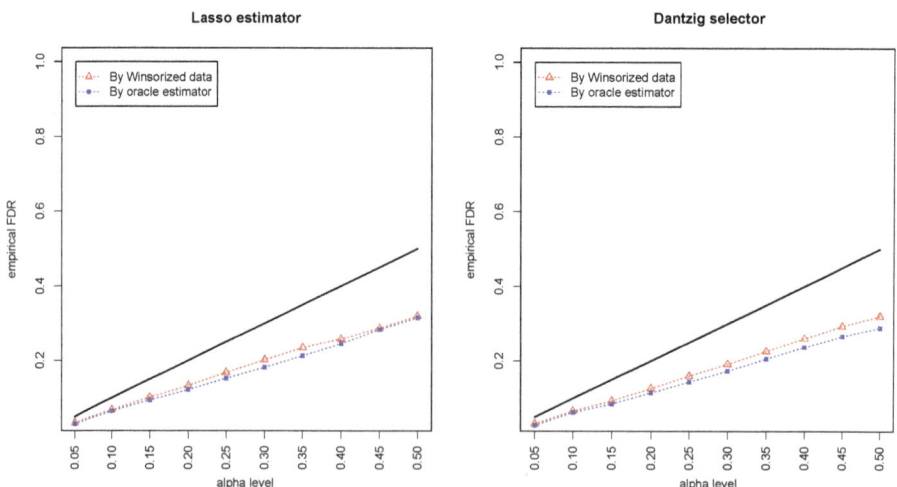

Figure 1. Empirical false discovery rates (eFDRs) by oracle data and Winsorized imputations under the band graph setting. The *x*-axis represents the desired FDR levels from 0.05 to 0.5, and the solid line is $y = x$.

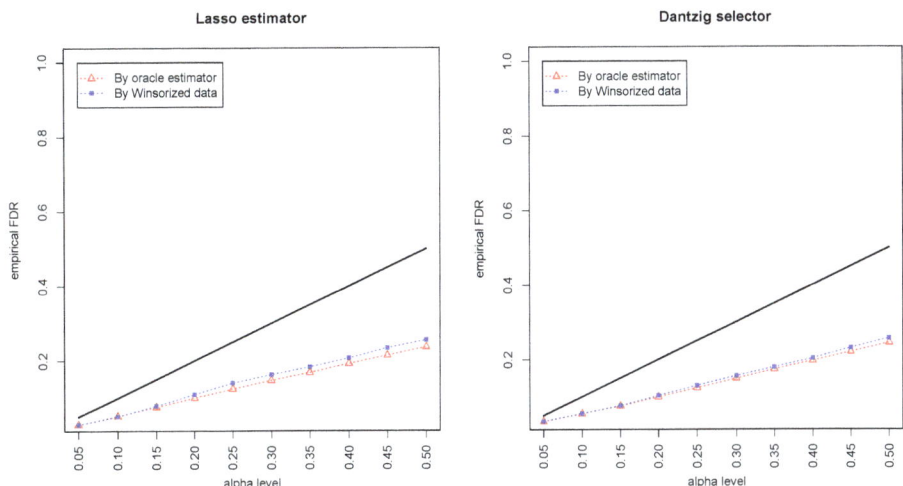

Figure 2. Empirical FDRs (eFDRs) by oracle data and Winsorized imputations under the Erdős–Rényi (ER) graph setting. The x-axis represents the desired FDR levels from 0.05 to 0.5, and the solid line is $y = x$.

Figures 3 and 4 summarized the statistical power of the test for the band graph and ER graph. As can be seen, the power for ER graph is substantially lower than the band graph, indicating that the complexity and denseness of the underlying differential network may significantly decrease the power of our test. The test based on oracle data performs slightly better than the imputed data, which is due to the loss of information during Winsorized imputation. Similar as we observed from Figures 1 and 2, the lasso estimator works almost equally well as Dantzig selector.

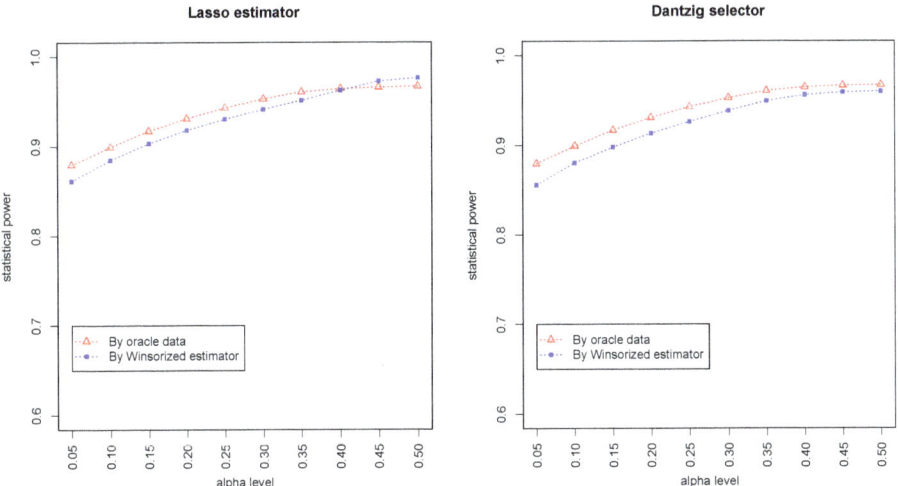

Figure 3. Statistical powers by oracle data and Winsorized imputations under the band graph setting. The x-axis represents the desired FDR levels from 0.05 to 0.5.

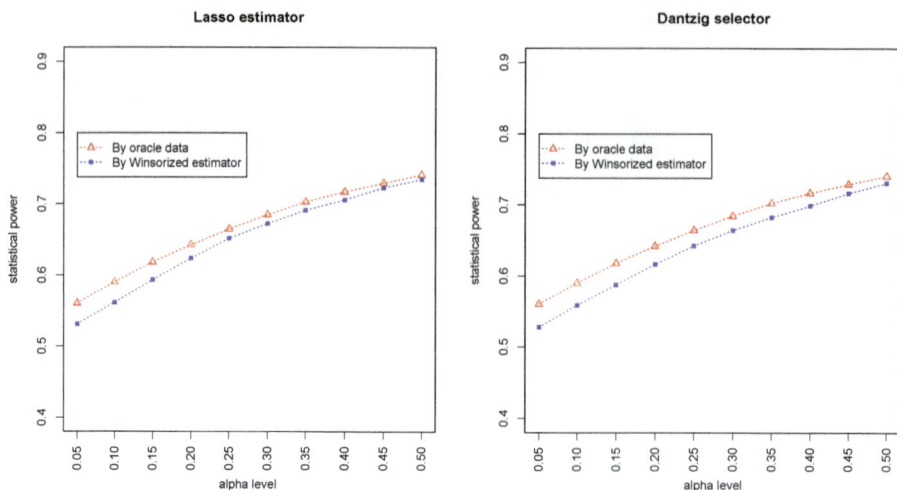

Figure 4. Statistical powers by oracle data and Winsorized estimator under the ER graph setting. The *x*-axis represents the desired FDR levels from 0.05 to 0.5.

In addition, we compared the proposed test with a direct estimator, recently developed by Zhang (2019) [8]. The direct estimator is a rank-based estimator and can be solved by a parametric simplex algorithm. We simulated the data from the Erdős–Rényi (ER) graph with different sample sizes ($n = 25, 50, 100, 150$) and numbers of dimensions ($p = 40, 60, 90, 120$). As the direct estimator does not control the false discovery rate, we set the FDR level at 0.05 for our proposed test. Figure 5 summarized the empirical FDR and statistical power under different sample sizes (with dimension fixed at 100) and different dimensions (with sample size fixed at 100). It can be seen that the two methods have comparable performance and our proposed test achieves lower FDR but slightly lower statistical power. However, it is noteworthy that the direct estimator is computationally expensive and becomes impractical when the dimensions exceed 150. Table 1 summarized the running time of the two methods, where it can be seen that our test is much faster than the direct estimator, especially for relatively high dimensions. For instance, when $p = 120$, the direct estimator takes hours while our test takes less than 10 seconds. As the core part of the proposed algorithm is the estimation of regression coefficients, the time complexity is the same as the linear regression. For instance, with LASSO and $p > n$, the time complexity is $O(np^2)$, while the direct estimator by Zhang (2019) has a time complexity $O(np^4)$.

Table 1. Running time of the proposed test and direct estimator (in seconds).

n/p	40	60	90	120
25	0.88 (7.0)	1.59 (110)	3.79 (1936)	6.49 (23,066)
50	1.19 (7.7)	1.93 (127)	4.15 (1973)	6.83 (23,119)
100	1.87 (9.1)	2.61 (146)	5.00 (2016)	7.80 (23,153)
200	2.11 (11)	3.04 (165)	6.22 (2055)	9.61 (23,201)

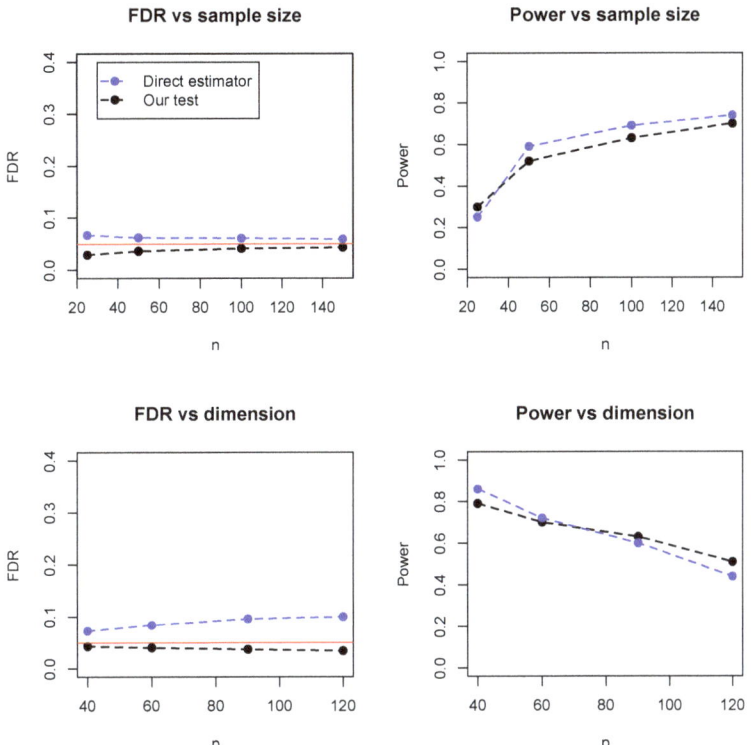

Figure 5. Comparison of the proposed test and direct estimator by Zhang (2019), in terms of empirical FDR and statistical power under different sample sizes and dimensions.

4. A Genomic Application

In this part, we applied the proposed test to the Cancer Genome Atlas data (TCGA, [9]) to study the different roles of the cell cycle pathway in the two subtypes of breast cancer including luminal A subtype and basal-like subtype. The cell cycle pathway is known to play a critical role in the initiation and progression of many human cancers including breast cancer and ovarian cancer [10,11]. For instance, the cell cycle pathway provided by KEGG (Kyoto Encyclopedia of Genes and Genomes, [12]) contains 128 important genes that co-regulate cell proliferation, including *ATM*, *RB1*, *CCNE1*, and *MYC*. Abnormal regulation among these genes may cause the over-proliferation of cells and an accumulation of tumor cell numbers [11].

The transcriptome profiling data for breast cancer were downloaded through the Genomic Data Commons portal [13] in January 2017. The expression level of each gene was quantified by the count of reads mapped to the gene. The quantifications were done by software *HTSeq* of version 0.9.1 [14]. In our analysis, we excluded 43 subjects including 12 male subjects and 31 subjects with >1% missing values. In addition, we removed the effects due to different age groups and batches using a median-matching and variance-matching strategy [10,15,16]. For example, the batch effect can be removed in the following way:

$$g^*_{ijk} = M_i + (g_{ijk} - M_{ij})\frac{\hat{\sigma}_{g_i}}{\hat{\sigma}_{g_{ij}}},$$

where g_{ijk} refers to the expression value for gene i from sample k in batch j ($j = 1, 2, ..., J; k = 1, 2, ..., n_j$), M_{ij} represents the median of $g_{ij} = (g_{ij1}, ..., g_{ijn_j})$, M_i refers to the median of $g_i = (g_{i1}, ..., g_{iJ})$, $\hat{\sigma}_{g_i}$ and $\hat{\sigma}_{g_{ij}}$ stand for the standard deviations of g_i and g_{ij}, respectively.

The remaining 959 breast cancer samples were further classified into five subtypes according to two molecular signatures, namely *PAM50* [17] and *SCMOD2* [18]. The two classifications were implemented separately using R package *genefu* [19] and we obtained 530 subjects with concordant classification by two classifiers. The resulting set contains 221 subjects in the luminal A group, 119 in the luminal B group, 74 in the her2-enriched group, 105 in the basal-like group, and 11 in the normal-like group. For illustration purposes, we conducted two pairwise comparisons (1) Luminal A vs basal-like and (2) Luminal B vs basal-like.

To balance the bias and variance, we choose the same truncation parameter in Winsorized imputation as in our simulation study

$$\delta_n^{(k)} = \frac{1}{4n_k^{1/4}\sqrt{\pi \log n_k}},$$

where $k \in \{1, 2\}$, $n_1 = 221$, $n_2 = 105$. The proposed test based on the Winsorized estimator was then conducted for each gene pair with different FDR cutoffs. Figures 6 and 7 summarized all the identified differential edges under FDR levels $\alpha = 0.05, 0.10, 0.15, 0.20$, with all isolated genes being removed. Our results suggested a list of important genes that play different roles in different breast cancer subtypes. For instance, in Figure 6, genes *CCNB1* and *PRKDC* contribute to several differential edges. According to recent studies, gene *CCNB1* is a prognostic biomarker for certain subtypes of breast cancer and it is closely associated with hormone therapy resistance [20]. It has also been reported in the literature that the *PRKDC* regulates chemosensitivity and is a potential prognostic and predictive marker of response to adjuvant chemotherapy in breast cancer patients [21]. Our findings about several other genes including *CHEK2* and *CDC7* also confirmed some existing reports [22,23]. As we observed from the two examples, as the desired FDR level increases, the resulting differential network tends to be denser and denser (Figure 8 showed the correlation between FDR and the number of differential edges). In practice, users should consider the trade-off between the accuracy (FDR) and number of new hypotheses (number of differential edges) and choose an appropriate FDR [24].

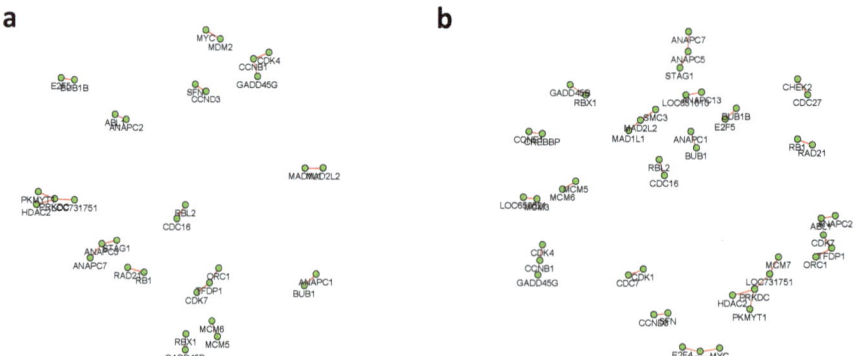

Figure 6. *Cont.*

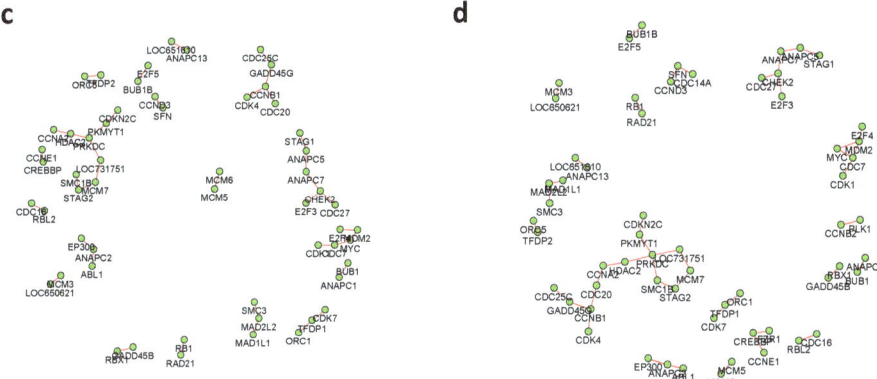

Figure 6. The inferred differential networks between the LumA and Basal-like subtypes under different desired FDR levels: (**a**) 0.05; (**b**) 0.10; (**c**) 0.15; (**d**) 0.20, with all isolated genes being removed. Each connection in the network represents an identified differential edge.

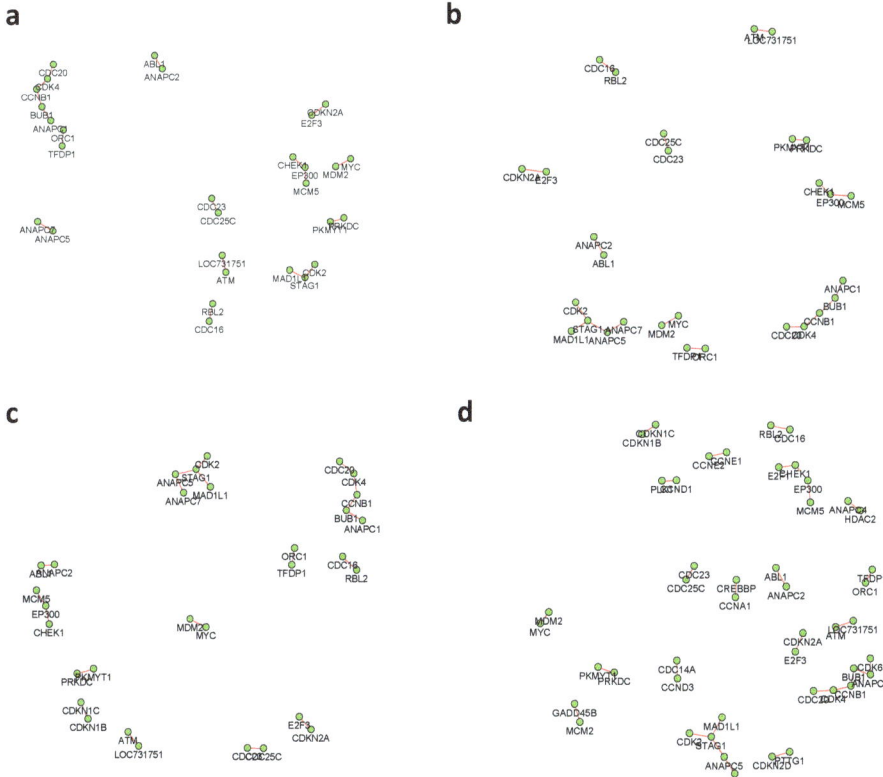

Figure 7. The inferred differential networks between the LumB and Basal-like subtypes under different desired FDR levels: (**a**) 0.05; (**b**) 0.10; (**c**) 0.15; (**d**) 0.20, with all isolated genes being removed. Each connection in the network represents an identified differential edge.

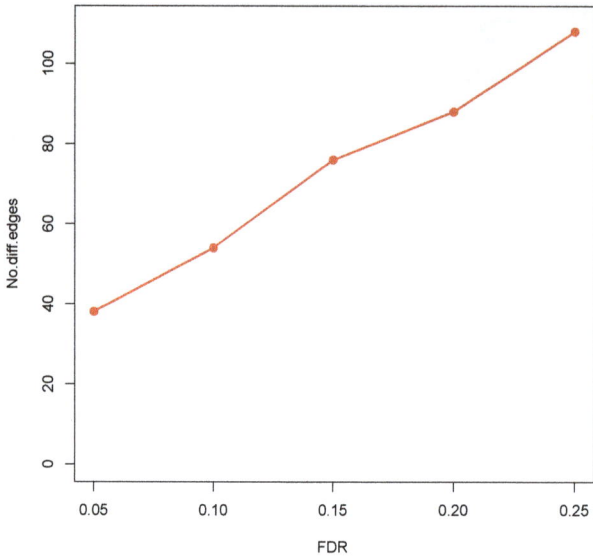

Figure 8. Desired FDR level against the number of differential edges.

5. Discussion

Detecting the differential substructure on multiple graphical models is a fundamental and challenging problem in statistics. Liu (2017) studied the problem under the Gaussian framework and introduced an elegant hierarchical test based on the estimation of single GGM. Unlike most existing methods, Liu's approach asymptotically controlled the false discovery rate at a nominal level, which guarantees the quality of the estimated differential network. In this work, we further extended Liu's test to a more flexible semiparametric framework, namely the nonparanormal graphical models. Our test is built upon a Winsorized estimator of the unknown transformation functions and it enjoys similar asymptotic properties as its oracle counterpart does.

Although the new test holds great promise in many applications such as genetic network modeling, it has some practical limitations. First, as we see from the theoretical derivation, the good performance of the test relied on the sparsity assumption on the differential network. Although the sparsity assumption is reasonable in many cases, it still could be violated in some applications. For instance, some genetic pathways may exhibit a global change of gene–gene regulations between different phenotypes. When the differential network is dense or locally dense, the method may fail to control the FDR. To solve the problem, a new test needs to be defined to evaluate the level of the sparseness of the change between two conditions. However, there is still a gap on the literature of this topic.

Second, one key assumption in NPNGMs is that the transformed variables follow a joint Gaussian distribution. This assumption also needs to be checked in real-world applications. Under low dimensions, one can employ some popular normality tests, including the Anderson–Darling test and Shapiro–Wilk test, on the imputed data or other normal scores. However, most of these tests fail to detect non-normality for high-dimension data. The normality test under high dimension is still an open and challenging problem and we left it for future research.

It is also noteworthy to mention that the new test relied on an accurate estimator for the coefficients β. Motivated by [6], we chose two popular estimators including lasso estimator and Dantzig selector based on the adjusted Spearman's rank, which satisfies Condition (7). In fact, some other estimators also satisfy the conditions, for instance, the rank-based adaptive lasso [6,25] and square-root lasso estimator [6,26]. These estimators can also be incorporated into our testing framework.

6. Conclusions

We have introduced a novel statistical test to detect the structural difference between the two nonparanormal graphical models. The proposed test dropped the Gaussian assumption and can be potentially applied to many non-Gaussian data for differential network analysis. For instance, some digital gene expression data (e.g., RNA-seq data) do not follow Gaussian distribution even after log transformation or other variance-stabilizing transformations. In such cases, one can model the data with a nonparanormal graphical model and apply our test to find differential edges between two or multiple phenotypic conditions. The proposed test may also be used to detect the difference between normal and disease populations in the brain connectivity network.

Funding: This research received no external funding.

Conflicts of Interest: The author declares no conflict of interest.

Abbreviations

The following abbreviations are used in this manuscript:

FDR   False discovery rate
NPNGM   Nonparanormal graphical model
GGM   Gaussian graphical model
TCGA   The Cancer Genome Atlas

Appendix A. Proof of Equation (6)

Define the estimated residuals based on Winsorized estimator as:

$$\epsilon_{mj}^* = X_{mj}^* - \bar{X}_j^* - (X_{m,-j}^* - \bar{X}_{-j}^*)\hat{\beta}_j,$$

where $\bar{X}_j^* = 1/n \sum_{m=1}^n X_{mj}^*$ and $\bar{X}_{-j}^* = 1/n \sum_{m=1}^n X_{m,-j}^*$. The choice of $\hat{\beta}_j$ must satisfy the following two conditions:

$$\|\hat{\beta}_j - \beta_j\|_{\ell_1} = O_p(a_n),$$

$$\min\left\{\lambda_{\max}^{1/2}(\Sigma)\|\hat{\beta}_j - \beta_j\|_{\ell_2}, \max_{1 \le j \le p} \sqrt{(\hat{\beta}_j - \beta_j)^T \hat{\Sigma}_{-j,-j}(\hat{\beta}_j - \beta_j)}\right\} = O_p(b_n),$$

where $a_n^{(k)} = o(\sqrt{\log p/n_k})$, and $b_n^{(k)} = o(n_k^{-1/4})$.

It is noteworthy to mention that the conditions above are slightly different from the conditions in [4] due to the different convergence rates by oracle data and imputed data. The conditions above can be satisfied by the rank-based estimators introduced in [6], e.g., rank-based lasso estimator or rank-based Dantzig selector. By letting $\tilde{\epsilon}_{mj} = \epsilon_{mj} - \bar{\epsilon}_i$, we have:

$$\epsilon_{mi}^* \epsilon_{mj}^* = \tilde{\epsilon}_{mi}\tilde{\epsilon}_{mj} - \tilde{\epsilon}_{mi}\{(X_{m,-j}^* - \bar{X}_{-j}^*)\hat{\beta}_j - (X_{m,-j} - \bar{X}_{-j})\beta_j\} \quad \text{(A1)}$$

$$- \tilde{\epsilon}_{mj}\{(X_{m,-i}^* - \bar{X}_{-i}^*)\hat{\beta}_i - (X_{m,-i} - \bar{X}_{-i})\beta_i\} \quad \text{(A2)}$$

$$+ \{\hat{\beta}_i^T(X_{m,-i}^* - \bar{X}_{-i}^*)^T(X_{m,-j}^* - \bar{X}_{-j}^*)\hat{\beta}_j - \beta_i^T(X_{m,-i} - \bar{X}_{-i})^T(X_{m,-j} - \bar{X}_{-j})\beta_j\}. \quad \text{(A3)}$$

First, for term (A3), we have:

$$|\frac{1}{n}\sum_{m=1}^{n}\{\hat{\boldsymbol{\beta}}_i^T(\mathbf{X}_{m,-i}^*-\bar{\mathbf{X}}_{-i}^*)^T(\mathbf{X}_{m,-j}^*-\bar{\mathbf{X}}_{-j}^*)\hat{\boldsymbol{\beta}}_j - \boldsymbol{\beta}_i^T(\mathbf{X}_{m,-i}-\bar{\mathbf{X}}_{-i})^T(\mathbf{X}_{m,-j}-\bar{\mathbf{X}}_{-j})\boldsymbol{\beta}_j\}|$$
$$=|\hat{\boldsymbol{\beta}}_i^T(\tilde{\boldsymbol{\Sigma}}_{-i,-j}-\hat{\boldsymbol{\Sigma}}_{-i,-j})\hat{\boldsymbol{\beta}}_j + (\hat{\boldsymbol{\beta}}_i-\boldsymbol{\beta}_i)^T(\hat{\boldsymbol{\Sigma}}_{-i,-j}-\boldsymbol{\Sigma}_{-i,-j})(\hat{\boldsymbol{\beta}}_j-\boldsymbol{\beta}_j) + (\hat{\boldsymbol{\beta}}_i-\boldsymbol{\beta}_i)^T\boldsymbol{\Sigma}_{-i,-j}(\hat{\boldsymbol{\beta}}_j-\boldsymbol{\beta}_j)|$$
$$\leq \max_{i,j}|\hat{\boldsymbol{\beta}}_i^T(\tilde{\boldsymbol{\Sigma}}_{-i,-j}-\hat{\boldsymbol{\Sigma}}_{-i,-j})\hat{\boldsymbol{\beta}}_j| + \max_{i,j}|(\hat{\boldsymbol{\beta}}_i-\boldsymbol{\beta}_i)^T(\hat{\boldsymbol{\Sigma}}_{-i,-j}-\boldsymbol{\Sigma}_{-i,-j})(\hat{\boldsymbol{\beta}}_j-\boldsymbol{\beta}_j)| + \max_{i,j}|(\hat{\boldsymbol{\beta}}_i-\boldsymbol{\beta}_i)^T\boldsymbol{\Sigma}_{-i,-j}(\hat{\boldsymbol{\beta}}_j-\boldsymbol{\beta}_j)|,$$

where the last term can be bounded as follows:

$$\max_{i,j}|(\hat{\boldsymbol{\beta}}_i-\boldsymbol{\beta}_i)^T\boldsymbol{\Sigma}_{-i,-j}(\hat{\boldsymbol{\beta}}_j-\boldsymbol{\beta}_j)| = O_p(\lambda_{\max}(\boldsymbol{\Sigma})\max_{1\leq i\leq p}\|\hat{\boldsymbol{\beta}}_i-\boldsymbol{\beta}_i\|_{\ell_2}^2) = O_p(b_n^2).$$

It is not hard to show that:

$$\|\hat{\boldsymbol{\Sigma}}-\boldsymbol{\Sigma}\|_\infty = O_p\left(\sqrt{\frac{\log p}{n}}\right),$$

therefore, the second term can also be bounded

$$\max_{i,j}|(\hat{\boldsymbol{\beta}}_i-\boldsymbol{\beta}_i)^T(\hat{\boldsymbol{\Sigma}}_{-i,-j}-\boldsymbol{\Sigma}_{-i,-j})(\hat{\boldsymbol{\beta}}_j-\boldsymbol{\beta}_j)| = O_p\left(a_n^2\sqrt{\frac{\log p}{n}}\right).$$

Under some mild regularity conditions (stated in [6]), we have

$$\|\tilde{\boldsymbol{\Sigma}}-\hat{\boldsymbol{\Sigma}}\|_\infty = O_p\left(\sqrt{\frac{\log p\log^2 n}{n^{1/2}}}\right),$$

thus under the condition that $\max_{i,j}|\beta_{i,j}| \leq C_1$ and $\lambda_{\min}(\boldsymbol{\Sigma}) = o((\log p/n)^{\frac{3}{4}})$, the first term can be bounded as follows:

$$\max_{i,j}|\hat{\boldsymbol{\beta}}_i^T(\tilde{\boldsymbol{\Sigma}}_{-i,-j}-\hat{\boldsymbol{\Sigma}}_{-i,-j})\hat{\boldsymbol{\beta}}_j|$$
$$\leq \max_{i,j}|\boldsymbol{\beta}_i^T(\tilde{\boldsymbol{\Sigma}}_{-i,-j}-\hat{\boldsymbol{\Sigma}}_{-i,-j})\boldsymbol{\beta}_j| + \max_{i,j}|(\hat{\boldsymbol{\beta}}_i-\boldsymbol{\beta}_i)^T(\tilde{\boldsymbol{\Sigma}}_{-i,-j}-\hat{\boldsymbol{\Sigma}}_{-i,-j})(\hat{\boldsymbol{\beta}}_j-\boldsymbol{\beta}_j)|$$
$$=O_p\left(\sqrt{\frac{\log^2 p\log^2 n}{n^{3/2}}} + a_n^2\sqrt{\frac{\log p\log^2 n}{n^{1/2}}}\right).$$

Combining the three terms above, we have

$$|\frac{1}{n}\sum_{m=1}^{n}\{\hat{\boldsymbol{\beta}}_i^T(\mathbf{X}_{m,-i}^*-\bar{\mathbf{X}}_{-i}^*)^T(\mathbf{X}_{m,-j}^*-\bar{\mathbf{X}}_{-j}^*)\hat{\boldsymbol{\beta}}_j - \boldsymbol{\beta}_i^T(\mathbf{X}_{m,-i}-\bar{\mathbf{X}}_{-i})^T(\mathbf{X}_{m,-j}-\bar{\mathbf{X}}_{-j})\boldsymbol{\beta}_j\}|$$
$$=O_p\left(\sqrt{\frac{\log^2 p\log^2 n}{n^{3/2}}} + a_n^2\sqrt{\frac{\log p\log^2 n}{n^{1/2}}} + b_n^2\right).$$

Next, we bound term (A1), which can be rewritten as:

$$\tilde{\epsilon}_{mi}(\mathbf{X}_{m,-j}-\bar{\mathbf{X}}_{-j})(\hat{\boldsymbol{\beta}}_j-\boldsymbol{\beta}_j) + \tilde{\epsilon}_{mi}\{(\mathbf{X}_{m,-j}-\bar{\mathbf{X}}_{-j}) - (\mathbf{X}_{m,-j}^*-\bar{\mathbf{X}}_{-j}^*)\}\hat{\boldsymbol{\beta}}_j,$$

where the first term can be further decomposed into two parts,

$$\tilde{\epsilon}_{mi}(\mathbf{X}_{m,-j}-\bar{\mathbf{X}}_{-j})(\hat{\boldsymbol{\beta}}_j-\boldsymbol{\beta}_j) = \tilde{\epsilon}_{mi}(X_{mi}-\bar{X}_i)(\hat{\beta}_{i,j}-\beta_{i,j})I\{i\neq j\} + \sum_{l\neq i,j}\tilde{\epsilon}_{mi}(X_{ml}-\bar{X}_l)(\hat{\beta}_{l,j}-\beta_{l,j}).$$

To bound $\sum_{l \neq i,j} \tilde{\epsilon}_{mi}(X_{ml} - \bar{X}_l)(\hat{\beta}_{l,j} - \beta_{l,j})$, we use the independence between ϵ_{mi} and $\mathbf{X}_{m,-i}$. It is easy to show that

$$\max_{l \neq i} |\frac{1}{n} \sum_{m=1}^{n} \tilde{\epsilon}_{mi}(X_{ml} - \bar{X}_l)| = O_p\left(\sqrt{\frac{\log p}{n}}\right),$$

which indicates that

$$\max_{i,j} |\frac{1}{n} \sum_{m=1}^{n} (\sum_{l \neq i,j} \tilde{\epsilon}_{mi}(X_{ml} - \bar{X}_l)(\hat{\beta}_{l,j} - \beta_{l,j}))| = O_p\left(a_n \sqrt{\frac{\log p}{n}}\right).$$

By the independence between ϵ_{mi} and $\mathbf{X}_m^* - \mathbf{X}_m$, it is not hard to show

$$|\frac{1}{n} \sum_{m=1}^{n} \tilde{\epsilon}_{mi}\{(\mathbf{X}_{m,-j} - \bar{\mathbf{X}}_{-j}) - (\mathbf{X}_{m,-j}^* - \bar{\mathbf{X}}_{-j}^*)\}\hat{\boldsymbol{\beta}}_j|$$

$$\leq |\frac{1}{n} \sum_{m=1}^{n} \tilde{\epsilon}_{mi}\{(\mathbf{X}_{m,-j} - \bar{\mathbf{X}}_{-j}) - (\mathbf{X}_{m,-j}^* - \bar{\mathbf{X}}_{-j}^*)\}\boldsymbol{\beta}_j| + |\frac{1}{n} \sum_{m=1}^{n} \tilde{\epsilon}_{mi}\{(\mathbf{X}_{m,-j} - \bar{\mathbf{X}}_{-j}) - (\mathbf{X}_{m,-j}^* - \bar{\mathbf{X}}_{-j}^*)\}(\boldsymbol{\beta}_j - \hat{\boldsymbol{\beta}}_j)|$$

$$= O_p\left(\frac{\log p}{n} + a_n \sqrt{\frac{\log p}{n}}\right).$$

Combing term (A1) and term (A2), we have

$$\frac{1}{n} \sum_{m=1}^{n} \tilde{\epsilon}_{mi}\{(\mathbf{X}_{m,-j}^* - \bar{\mathbf{X}}_{-j}^*)\hat{\boldsymbol{\beta}}_j - (\mathbf{X}_{m,-j} - \bar{\mathbf{X}}_{-j})\}\boldsymbol{\beta}_j = \frac{1}{n} \sum_{m=1}^{n} \tilde{\epsilon}_{mi}(X_{mi} - \bar{X}_i)(\hat{\beta}_{i,j} - \beta_{i,j})I\{i \neq j\}$$

$$+ O_p\left(\frac{\log p}{n} + a_n \sqrt{\frac{\log p}{n}}\right).$$

Summarizing all the results above, by Equations (22) and (23) of Liu (2013), we have

$$\frac{1}{n} \sum_{m=1}^{n} \epsilon_{mi}^* \epsilon_{mj}^* = \frac{1}{n} \sum_{m=1}^{n} \tilde{\epsilon}_{mi} \tilde{\epsilon}_{mj} - \frac{1}{n} \sum_{m=1}^{n} \tilde{\epsilon}_{mi}(X_{mi} - \bar{X}_i)(\hat{\beta}_{i,j} - \beta_{i,j})I\{i \neq j\}$$

$$- \frac{1}{n} \sum_{m=1}^{n} \tilde{\epsilon}_{mj}(X_{mj} - \bar{X}_j)(\hat{\beta}_{j,i} - \beta_{j,i})I\{i \neq j\}$$

$$+ O_p\left(\sqrt{\frac{\log^2 p \log^2 n}{n^{3/2}}} + a_n^2 \sqrt{\frac{\log p \log^2 n}{n^{1/2}}} + a_n \sqrt{\frac{\log p}{n}} + b_n^2\right).$$

As $\frac{1}{n} \sum_{m=1}^{n} \tilde{\epsilon}_{mi}(X_{mi} - \bar{X}_i) = \frac{1}{n} \sum_{m=1}^{n} \tilde{\epsilon}_{mi}^2 + \frac{1}{n}\tilde{\epsilon}_{mi}(\mathbf{X}_{m,-i} - \bar{\mathbf{X}}_{-i})\boldsymbol{\beta}_i$, and $\text{Var}(\mathbf{X}_{m,-i}\boldsymbol{\beta}_i) = (\sigma_{ii}\omega_{ii} - 1)/\omega_{ii} \leq C$, we have

$$\max_{1 \leq i \leq p} |\frac{1}{n} \sum_{m=1}^{n} \tilde{\epsilon}_{mi}(\mathbf{X}_{m,-i} - \bar{\mathbf{X}}_{-i})\boldsymbol{\beta}| = O_p\left(\sqrt{\frac{\log p}{n}}\right),$$

therefore

$$\frac{1}{n} \sum_{m=1}^{n} \tilde{\epsilon}_{mi}(X_{mi} - \bar{X}_i) = \frac{1}{n} \sum_{m=1}^{n} \tilde{\epsilon}_{mi}^2 + O_p(\log p/n)$$

$$= \frac{1}{n} \sum_{m=1}^{n} \epsilon_{mi}^{*2} + O_p\left(\sqrt{\frac{\log^2 p \log^2 n}{n^{3/2}}} + a_n^2 \sqrt{\frac{\log p \log^2 n}{n^{1/2}}} + a_n \sqrt{\frac{\log p}{n}} + b_n^2\right),$$

$$\frac{1}{n}\sum_{m=1}^{n}\epsilon_{mi}^{*}\epsilon_{mj}^{*} = \frac{1}{n}\sum_{m=1}^{n}\tilde{\epsilon}_{mi}\tilde{\epsilon}_{mj} - \frac{1}{n}\sum_{m=1}^{n}\epsilon_{mi}^{*2}(\hat{\beta}_{i,j} - \beta_{i,j})I\{i \neq j\} - \frac{1}{n}\sum_{m=1}^{n}\epsilon_{mj}^{*2}(\hat{\beta}_{j,i} - \beta_{j,i})I\{i \neq j\}$$
$$+ O_p\left(\sqrt{\frac{\log^2 p \log^2 n}{n^{3/2}}} + a_n^2\sqrt{\frac{\log p \log^2 n}{n^{1/2}}} + a_n\sqrt{\frac{\log p}{n}} + b_n^2\right).$$

In addition

$$\frac{1}{n}\sum_{m=1}^{n}\epsilon_{mi}^{*2} = \frac{1}{n}\sum_{m=1}^{n}\tilde{\epsilon}_{mi}^{2} + O_p\left(\sqrt{\frac{\log^2 p \log^2 n}{n^{3/2}}} + a_n^2\sqrt{\frac{\log p \log^2 n}{n^{1/2}}} + a_n\sqrt{\frac{\log p}{n}} + b_n^2\right).$$

Equation (6) follows immediately by central limit theorem.

References

1. Guo, J.; Levina, E.; Michailidis, G.; Cai, T.T. Joint estimation of multiple graphical models. *Biometrika* **2015**, *98*, 1–15. [CrossRef]
2. Danaher, P.; Wang, P.; Witten, D.M. The joint graphical lasso for inverse covariance estimation across multiple classes. *J. R. Stat. Soc. Ser. B* **2014**, *76*, 373–397. [CrossRef]
3. Zhao, S.; Cai, T.T.; Li, H. Direct estimation of differential networks. *Biometrika* **2014**, *101*, 253–268. [CrossRef]
4. Liu, W. Structural similarity and difference testing on multiple sparse Gaussian graphical models. *Ann. Stat.* **2017**, *45*, 2680–2707. [CrossRef]
5. Liu, H.; Lafferty, J.; Wasserman, L. The nonparanormal: Semiparametric estimation of high dimensional undirected graphs. *J. Mach. Learn. Res.* **2009**, *10*, 2295–2328.
6. Xue, L.; Zou, H. The nonparanormal: Semiparametric estimation of high dimensional undirected graphs. *Ann. Stat.* **2012**, *40*, 2541–2571. [CrossRef]
7. Efron, B. Correlation and large-scale simultaneous significance testing. *J. Am. Stat. Assoc.* **2007**, *102*, 93–103. [CrossRef]
8. Zhang, Q. Direct estimation of differential network under high-dimensional nonparanormal graphical models. *Can. J. Stat.* **2019**, *48*, 1–17. [CrossRef]
9. Available online: https://cancergenome.nih.gov (accessed on 9 December 2019).
10. Zhang, Q.; Burdette, J.; Wang, J.-P. Integrative network analysis of tcga data for ovarian cancer. *BMC Syst. Biol.* **2014**, *8*, 1–18. [CrossRef]
11. The Cancer Genome Atlas Network. Comprehensive molecular portraits of human breast tumours. *Nature* **2012**, *490*, 61–70. [CrossRef]
12. Available online: http://www.genome.jp/kegg/pathway (accessed on 9 December 2019).
13. Available online: https://gdc.cancer.gov (accessed on 9 December 2019).
14. Anders, S.; Pyl, P.T.; Huber, W. HTSeq—A Python framework to work with high-throughput sequencing data. *Bioinformatics* **2015**, *31*, 166–169. [CrossRef] [PubMed]
15. Hsu, F.; Serpedin, E.; Hsiao, T.; Bishop, A.; Dougherty, E.; Chen, Y. Reducing confounding and suppression effects in tcga data: An integrated analysis of chemotherapy response in ovarian cancer. *BMC Genom.* **2012**, *13*, S13. [CrossRef] [PubMed]
16. Zhang, Q. A powerful nonparametric method for detecting differentially co-expressed genes: Distance correlation screening and edge-count test. *BMC Syst. Biol.* **2018**, *12*, 58. [CrossRef] [PubMed]
17. Liu, M.C.; Pitcher, B.N.; Mardis, E.R.; Davies, S.R.; Friedman, P.N.; Snider, J.E.; Vickery, T.L.; Reed, J.P.; DeSchryver, K.; Singh, B.; et al. PAM50 gene signatures and breast cancer prognosis with adjuvant anthracycline- and taxane-based chemotherapy: Correlative analysis of C9741. *Breast Cancer* **2016**, *2*, 15023. [CrossRef]
18. Haibe-Kains, B.; Desmedt, C.; Loi, S.; Culhane, A.C.; Bontempi, G.; Quackenbush, J.; Sotiriou, C. A three-gene model to robustly identify breast cancer molecular subtypes. *J. Natl. Cancer Inst.* **2012**, *104*, 311–325. [CrossRef]
19. Gendoo, D.M.; Ratanasirigulchai, N.; Schroder, M.S.; Pare, L.; Parker, J.S.; Prat, A.; Haibe-Kains, B. Genefu: An R/Bioconductor package for computation of gene expression-based signatures in breast cancer. *Bioinformatics* **2016**, *32*, 1097–1099. [CrossRef]

20. Ding, K.; Li, W.; Zou, Z.; Zou, X.; Wang, C. CCNB1 is a prognostic biomarker for ER+ breast cancer. *Med. Hypothesis* **2014**, *83*, 359–364. [CrossRef]
21. Sun, G.; Yang, L.; Dong, C.; Ma, B.; Shan, M.; Ma, B. PRKDC regulates chemosensitivity and is a potential prognostic and predictive marker of response to adjuvant chemotherapy in breast cancer patients. *Oncol. Rep.* **2017**, *37*, 3536–3542. [CrossRef]
22. Huggett, M.; Tudzarova, S.; Proctor, I.; Loddo, M.; Keane, M.G.; Stoeber, K.; Williams, G.H.; Pereira, S.P. Cdc7 is a potent anti-cancer target in pancreatic cancer due to abrogation of the DNA origin activation checkpoint. *Oncotargets* **2016**, *7*, 18495–18407. [CrossRef]
23. Desrichard, A.; Bidet, Y.; Uhrhammer, N.; Bignon, Y. CHEK2 contribution to hereditary breast cancer in non-BRCAfamilies. *Breast Cancer Res.* **2011**, *13*, R119. [CrossRef]
24. Liu, W. Gaussian graphical model estimation with false discovery rate control. *Ann. Stat.* **2013**, *41*, 2948–2978. [CrossRef]
25. Zou, H. The Adaptive Lasso and Its Oracle Properties. *J. Am. Stat. Assoc.* **2006**, *101*, 1418–1429. [CrossRef]
26. Belloni, A.; Chernozhukov, V.; Wang, L. Square-root lasso: Pivotal recovery of sparse signals via conic programming. *Biometrika* **2011**, *98*, 791–806. [CrossRef]

© 2020 by the author. Licensee MDPI, Basel, Switzerland. This article is an open access article distributed under the terms and conditions of the Creative Commons Attribution (CC BY) license (http://creativecommons.org/licenses/by/4.0/).

Article

Detection of Differentially Methylated Regions Using Bayes Factor for Ordinal Group Responses

Fengjiao Dunbar [1], Hongyan Xu [2], Duchwan Ryu [3], Santu Ghosh [2], Huidong Shi [4] and Varghese George [2,*]

1. Genomics Research Center, AbbVie, North Chicago, IL 60064, USA; fengjiao.dunbar@abbvie.com
2. Department of Population Health Sciences, Augusta University, Augusta, GA 30912, USA; hxu@augusta.edu (H.X.); sghosh@augusta.edu (S.G.)
3. Department of Statistics and Actuarial Science, Northern Illinois University, DeKalb, IL 60178, USA; dryu@niu.edu
4. Georgia Cancer Center, Augusta University, Augusta, GA 30912, USA; hshi@augusta.edu
* Correspondence: vgeorge@augusta.edu; Tel.: +1-706-721-0801

Received: 19 July 2019; Accepted: 15 September 2019; Published: 17 September 2019

Abstract: Researchers in genomics are increasingly interested in epigenetic factors such as DNA methylation, because they play an important role in regulating gene expression without changes in the DNA sequence. There have been significant advances in developing statistical methods to detect differentially methylated regions (DMRs) associated with binary disease status. Most of these methods are being developed for detecting differential methylation rates between cases and controls. We consider multiple severity levels of disease, and develop a Bayesian statistical method to detect the region with increasing (or decreasing) methylation rates as the disease severity increases. Patients are classified into more than two groups, based on the disease severity (e.g., stages of cancer), and DMRs are detected by using moving windows along the genome. Within each window, the Bayes factor is calculated to test the hypothesis of monotonic increase in methylation rates corresponding to severity of the disease versus no difference. A mixed-effect model is used to incorporate the correlation of methylation rates of nearby CpG sites in the region. Results from extensive simulation indicate that our proposed method is statistically valid and reasonably powerful. We demonstrate our approach on a bisulfite sequencing dataset from a chronic lymphocytic leukemia (CLL) study.

Keywords: Bayes factor; Bayesian mixed-effect model; CpG sites; DNA methylation; Ordinal responses

1. Introduction

It is now widely accepted that cancer develops through a series of stages [1]. It starts from a very limited area, not invasive and metastatic at the early stage, then spreads to distant sites in the body, and becomes highly invasive and metastatic at the late stage. In addition, patient survival times are significantly reduced at the late stages. For example, the 5-year relative survival rate for lung cancer is 54% at a localized stage, and is reduced to 4% at the distant stage [2]. More than half of lung cancers are diagnosed at a distant stage, which indicates that early diagnosis of cancer is the main factor to enhance patient survival. Therefore, markers for early detection and proper classification of the tumor are extremely critical to improve life expectancy. Furthermore, identifying high-risk cancer patients at an early stage, would allow them to receive standard chemotherapy in advance.

DNA methylation has been found to be a marker for disease diagnosis, such as in cancer [3]. Significant progress has been made using DNA methylation differences to capture substantial information about the molecular and gene-regulatory states among biology subtypes, such as tumor and normal tissues [4].

In addition, DNA methylation can be used as a marker to differentiate disease severity, such as early and late stages in breast cancer [5], ovarian cancer [6] and prostate cancer [7]. Most of them have potential functions in inducing and suppressing cancer metastasis. Moreover, DNA methylation is associated with tumor size in colorectal cancer [8].Patients with higher methylation showed more frequent recurrence as compared with the low-methylation group, and shortened cancer-related survival and recurrence-free survival [8].

These findings show the critical importance of a better understanding of cancer progression and metastasis, which could help make better prediction of the clinical aggressiveness of cancer. Since DNA methylation is associated with disease severity, detecting differentially methylated regions (DMRs) can help understand cancer progression.

Most analyses are conducted by creating dichotomies based on biological subtypes, such as early and late cancer stages, and then detect DMRs by comparing the differences of DNA methylation rates between two groups [5–7]. However, when there are actually more than two groups, such approaches may lose information regarding multiple disease status, due to collapsing or ignoring clinically relevant subtypes, resulting in suboptimal clinical conclusions and decisions.

To use multiple disease status, it is possible to run multiple testing for the association between DNA methylation and multiple group responses, using the methods for two groups. Although we can simply run analysis for all pair-wise comparisons and combine the results, it is not trivial when considering the regional correlation of DMRs, and would increase the multiple testing burden.

Another possible method is the generalized linear model that includes indicator variables for different levels of disease status. This method has the advantage that it can adjust for covariates. However analysts are often faced with noisy estimates of category-specific regression coefficients, which can lead to unreasonable patterns in the regression coefficients corresponding to different levels of disease status, and it can reduce the power [9].

To improve the efficacy of an overall test, one can take advantage of the fact that cancer develops through a series of stages, or different levels of disease severity in general, and develop statistical methods that can incorporate the ordering of disease status. However, the widely used trend test is not an ideal method, because it requires scores or weights for different levels of disease status, which are generally unknown.

Here we propose a Bayesian approach and use the Bayes factor to test the association between methylation rates and disease severity. The proposed Bayes Factor Method (BFM) can incorporate monotonicity constraints, and find DMRs in which methylation rates increase (or decrease) as the diseases become more severe. Patients are classified into groups based on the disease severity (e.g., stages of cancer), and DMRs are detected by using moving windows along the genome. Within each window, the Bayes factor is calculated and is used to test the hypothesis of constant versus monotonic increase in methylation rates corresponding to the severity of the disease.

In addition, since DNA methylation rates have been shown to be correlated at nearby CpG sites with complicated correlation structure [10], a linear mixed-effect model is used to incorporate the correlation of methylation rates between and within CpG sites in the region.

2. Materials and Methods

2.1. Methods

Classical statistical inference under constrained parametric spaces has been addressed by many studies. Among them, Bartholomew [11] presented one of the first tests for K multinomial proportions with inequality constraints. He proposed a test of $H_0 : p_1 = p_2 = \ldots = p_K$ against the simple ordered $H_1 : p_1 \leq p_2 \leq \ldots \leq p_K$ with at least one strict inequality, where p_k ($k = 1, 2, \ldots, K$) represents the proportion the kth group. Under H_0, the maximum likelihood estimator of p_k is the overall sample proportion π_k. If the sample multinomial proportions satisfy $\pi_1 \leq \pi_2 \leq \ldots \leq \pi_K$, then the order-restricted ML estimator is $\hat{p}_k = \pi_k$. However, sometimes the sample proportions may not satisfy the ordering $\pi_1 \leq \pi_2 \leq \ldots \leq \pi_K$; in that case, calculation of the restricted maximum likelihood estimator (RMLE) is subject to arbitrary orderings of the parameters, and it requires specialized algorithms that are not easily generalizable [9].

Robertson and Wegman [12] proposed a likelihood ratio statistic for the inequality-constrained binomial problem, which compares parameters for independent samples from a single-parameter exponential family distribution. Before calculating the test statistic, they used the pool-adjacent-violators algorithm [13] to pool "out-of-order" categories for which $\pi_k > \pi_{k+1}$ until the resulting sample proportions are monotone increasing. The order-restricted ML estimators $\hat{p}_k$ become the adjusted sample proportions.

The idea of applying an isotonic transformation to the unconstrained parameter estimates motivated Dunson and Neelon [9] to create a Bayesian alternative approach for this problem, which has been adapted here. They proposed to use Bayes factors for assessing ordered trends, which are calculated based on the output from Gibbs sampling. The samples from the order-constrained model are derived by transforming samples draws from an unconstrained posterior density using an isotonic regression transformation. Next, we explain our proposed Bayes factor method (BFM).

Suppose m_{kij} is the count of methylated molecules at CpG site j of individual i in group k. We assume $m_{kij} \sim B(c_{kij}, p_{kij})$, where c_{kij} is the coverage, and p_{kij} is the true methylation rate at that particular site, with $k = 1, 2, \ldots, K$, $i = 1, 2, \ldots, n_k$ and $j = 1, 2, \ldots, m$.

Within each moving window along the genome, a mixed-effect model is considered to allow the correlation of methylation rates between and within CpG sites. The logit link function for the methylation rate p_{kij} is expressed by

$$\text{logit}(p_{kij}) = \mu_k + \nu_{0ki} + \nu_{1kij}, \tag{1}$$

where ν_{0ki} and ν_{1kij} are the random effects. The random effect $\nu_{0ki} \sim N(0, \sigma_{\nu_0}^2)$ is used to model the interindividual correlation of methylation rates within each CpG site, while the random effect $\nu_{1ki} = (\nu_{1ki1}, \nu_{1ki2}, \ldots, \nu_{1kim})^T \sim N(\mu_0, \Sigma)$, with $\mu_0 = (0, 0 \ldots 0)^T$ is used to model the correlation of methylation rates between CpG sites.

Here μ_k in (1) is the fixed effect for each group, representing the association between methylation rates and group responses. The strength and direction of the association is modeled by prior distribution $N(\mu_\mu, \sigma_\mu^2)$, which means the parameters of μ_μ and σ_μ^2 control the distribution of μ_k, and implies that all of the methylation rates are drawn from a common distribution. This brings the advantage of allowing for heterogeneity of effects across CpG sites, instead of just pooling information across CpG sites in a region. Pooling assumes that each CpG site in the region has same methylation rates, while BFM considers the methylation rates of each CpG sites to be a random quantity governed by a prior distribution.

With assigned hyperpriors $\mu_k \sim N(0, 1000^2)$, $\sigma_k^2 \sim IG(1, 100)$, $\sigma_{\nu_0}^2 \sim IG(1, 100)$ and $\Sigma^{-1} \sim \text{Wish}(I_m, m)$ for m CpG sites in the moving window. The posterior distribution of μ_k is based on the mixed-effect logistic model (1), and it is used to calculate the Bayes factor for comparing the two models, $M_0 : \mu_1 = \mu_2 = \ldots = \mu_K$, $M_1 : \mu_1 \leq \mu_2 \leq \ldots \leq \mu_K$ with at least one strict inequality, in order to see whether there is an ordered constraint of methylation rates corresponding to severity of the disease.

To calculate the Bayes factor, first we drew samples $\mu_1, \mu_2, \ldots, \mu_K$ from the posterior distribution by using Gibbs sampling. After that, an isotonic transformation is used to transform $\mu_1, \mu_2 \ldots \mu_K$ into $\widetilde{\mu}_1, \widetilde{\mu}_2, \ldots, \widetilde{\mu}_K$, with $\widetilde{\mu}_1 \leq \widetilde{\mu}_2 \leq \ldots \leq \widetilde{\mu}_K$ [8] by using the min-max formula for the isotonic transformation, given by,

$$\widetilde{\mu}_k = g_k(\mu) = \min_{t \in U_k} \max_{s \in L_k} \left(\frac{1'_{t-s+1} V_{[s:t]}^{-1} \mu_{[s:t]}}{1'_{t-s+1} V_{[s:t]}^{-1} 1_{t-s+1}} \right) \text{ for } j = 1, 2, \ldots, K, \tag{2}$$

where $V = \text{diag} V_1, \ldots, V_K$ denotes the posterior covariance matrix and the diagonal submatrix V_i, $i = 1, \ldots, k$, is the covariance matrix of the ith ordered group. It is estimated from the samples of the posterior density of μ. U_k and L_k denote subsets of $\{1, \ldots, K\}$ such that the ordering $\mu_{j'} \leq \mu_j$ for all $j' \in L_k$ and the ordering $\mu_{j'} \geq \mu_j$ for all $j' \in U_k$. Also samples $\mu_1^0, \mu_2^0, \ldots, \mu_K^0$ are drawn from the prior density and transformed into $\tilde{\mu}_1^0, \tilde{\mu}_2^0, \ldots, \tilde{\mu}_K^0$, with $\tilde{\mu}_1^0 \leq \tilde{\mu}_2^0 \leq \ldots \leq \tilde{\mu}_K^0$, by using the isotonic transformation in (2). The Bayes factor for each window (with moving windows along the genome) is given by,

$$BF = \frac{P(M_1|data)/P(M_1)}{P(M_0|data)/P(M_0)} = \frac{P(\tilde{\mu}_K > \tilde{\mu}_1)/P(\tilde{\mu}_K^0 > \tilde{\mu}_1^0)}{P(\tilde{\mu}_K = \tilde{\mu}_1)/P(\tilde{\mu}_K^0 = \tilde{\mu}_1^0)}$$

Please note that the isotonic transformation in (2) changes our hypotheses slightly, making the resulting Bayes Factor an approximation rather than exact [14]. The windows with highest value of the Bayes factor among all windows are used for evaluating DMRs.

Thus, the Bayes factor is the ratio of the marginal densities of the data under the two hypotheses, and it can be used to weigh evidence in favor of a hypothesis, by utilizing all the information contained in the full likelihood. Our proposed BFM can detect DMRs associated with disease severity, especially detecting DMRs with monotonically increasing or decreasing methylation rates, as the disease severity increase. It uses a mixed-effect model to not only adjust for correlation of methylation rates between CpG sites within each moving window but also correlations within CpG sites.

In addition, by adding covariates x_{ki} in the model (1), we can account for the effects of covariates that are associated with methylation rates, such as age [15] and gender [16].

To aid in the interpretation of the Bayes factor, Jeffreys [17] proposed the following rule of thumb: "When $3 < BF \leq 10$ the evidence is positive, when $10 < BF \leq 100$ the evidence is strong, and when $BF > 100$, the evidence is decisive". As Kass and Raftery [18] pointed out, these categories are not precise calibration, but rather a descriptive statement about the standards of evidence in scientific investigations.

2.2. Simulation Study of the Properties of BFM

Extensive simulation was conducted to study the statistical validity and power of BFM to detect DMRs. For simplicity, for each individual, we simulated one CpG island (genomic region with CpG sites) consisting of m equally spaced CpG sites, with only one DMR of length $r(<m)$ in the middle of the island. Further, we used equal sample size, N, for each of the K groups, and, we did not include any covariates.

Simulation Setup:

The goal here is to simulate methylation rate at each CpG site for each individual. This is achieved in two steps. In step 1, methylation data in the form of NGS short reads sequences were simulated for each CpG site, with correlated methylation status between CpG sites. We also assumed that methylation status at CpG sites among different sequences were independent, as expected in NGS data. In step 2, the individual methylation rates were calculated by summarizing the methylation status at each CpG site from the short read sequences.

The simulation details are described below:

First, we generated 100 NGS short reads using 100 pairs of random numbers {a, c} where a is the start point and c is the length of each short read sequence.

Then we used vector $\mathbf{Y} = (Y_{kis,a}, Y_{kis,a+1}, \ldots, Y_{kis,a+c-1})$ to define the methylation status for short read sequence s of individual i in group k, and generated $\mathbf{Y}$ from a multivariate Bernoulli distribution to allow for the correlation among the methylation rates. $P(\mathbf{Y} = \mathbf{y}) = P(y_{kisa}, y_{kis,a+1}, \ldots, y_{kis,a+c-1})$ of such a discrete random vector $\mathbf{Y}$ depends on 2^c probabilities, $p(0,0,\ldots,0), p(0,0,\ldots,1), \ldots, p(1,1,\ldots,1)$, specific to the different realizations of $\mathbf{Y}$. Considering the fact that if a vector $(Y_1, Y_2, \ldots, Y_p)$ follows p-variate Bernoulli distribution, the conditional distribution of $(Y_1, Y_2, \ldots, Y_r)$ $(r < p)$ given $(Y_{r+1}, Y_{r+2}, \ldots, Y_p)$ is also a multivariate Bernoulli distribution [18]. We can utilize this fact to reduce the dimensionality of the unconditional multivariate Bernoulli distribution.

Because of the correlation of methylation rates between CpG sites, we treated methylation status $Y_{kis,j}$ at each CpG site j on short read sequence s as a branching process, taking advantage of the property of multivariate Bernoulli distribution [19]. We assumed that, for CpG site j, branching probabilities were the same for each short read sequence of all individuals in group k. Thus, we defined the branching probability $p_{kj} = P(Y_{kis,j} = 1 | Y_{kis,j-1} = 1)$ as the probability of methylated sequence read at CpG site j, conditional on the methylated sequence read at CpG site j − 1 on the same short read sequence of the same individual. Similarly, we defined the branching probability $q_{kj} = P(Y_{kis,j} = 1 | Y_{kis,j-1} = 0)$ as the same probability, conditional on unmethylated sequence read at CpG site j − 1.

The methylation status $(Y_{kis,a}, Y_{kis,a+1}, \ldots, Y_{kis,a+c-1})$ were generated as follows:

For the first CpG site of the sequence, the methylation status y_{kisa} was generated from Bernoulli distribution $Bern(m_a)$, with $m_a = (p_{ka} + q_{ka})/2$.

The methylation status $y_{kis,j}$ for $j = a+1, \ldots, a+c-1$ was generated with $y_{kis,j} \sim Bern(p_{kj})$ if $y_{kis,j-1} = 1$ or $y_{kis,j} \sim Bern(q_{kj})$ if $y_{kis,j-1} = 0$.

After generating all the sequences at every CpG site for each individual, we calculated the total numbers of methylated and unmethylated short read sequences at CpG site j for individual i in group k, $\sum_s (y_{kis,j} = 1)$ and $\sum_s (y_{kis,j} = 0)$,. Then the methylation count and the sequencing coverage are given by $m_{kij} = \sum_s (y_{kis,j} = 1)$ and $c_{kij} = \sum_s (y_{kis,j} = 1) + \sum_s (y_{kis,j} = 0)$, respectively.

We generated one CpG region with 24 CpG sites for each individual, 6 of which (from site 10 to 15) constituting the DMR. We simulated four groups of severity levels, with sample size 50 in each group, and repeated it with sample size 100. The branching probabilities, p_{kj}, were pre-determined. Also, we chose $q_{kj} = p_{kj} - 0.2$. We also simulated two different scenarios of DMR patterns.

Under Scenario 1, we chose the probabilities, p_{kj}, to be symmetric around the middle of the DMR (CpG sites 12 and 13). The predetermined probabilities p_{kj} and their symmetric pattern under Scenario 1 are presented in Table 1.

Table 1. Conditional probabilities p_{kj} at each CpG site for simulation of BFM under Scenario 1.

Site	1	2	...	9	10	11	12	13	14	15	16	17	...	24
group 1	0.44	0.46	...	0.6	0.62	0.64	0.66	0.66	0.64	0.62	0.6	0.58	...	0.44
group 2	0.44	0.46	...	0.6	0.72	0.74	0.76	0.76	0.74	0.72	0.6	0.58	...	0.44
group 3	0.44	0.46	...	0.6	0.82	0.84	0.86	0.86	0.84	0.82	0.6	0.58	...	0.44
group 4	0.44	0.46	...	0.6	0.92	0.94	0.96	0.96	0.94	0.92	0.6	0.58	...	0.44

Under Scenario 2, we randomly chose the CpG sites with the peak values of p_{kj} within the simulated DMR (between sites 10 and 15), varying it for different individuals. Specifically, for each individual in each group, we first generated a random number r (between 10 and 15) for the location of the CpG site with the highest methylation, and then chose the branching probabilities p_{kj} to increase from 10 to r and then decrease from r to 15. The p_{kj} for the non-DMCs remained the same as in Scenario 1. The second scenario is a more realistic depiction of the real world. However, the results and conclusions should be the same under both situations.

3. Results

3.1. Simulation Results

A total of 1000 replicates were simulated. For each replicate, the Bayes factor was calculated for each moving window with window size of 6. Calculations were based on 3000 Gibbs samplers, with 1000 Gibbs samplers for the burn-in period. The results of simulation for both the scenarios are presented in Table 2. As expected, the results are very similar for both scenarios. The results of Scenario 1 are plotted in Figures 1 and 2. As evident from Table 2, following Jeffreys' rule, when the moving windows contain at least three of the six CpG sites, we have strong evidence of differential methylation when sample size of 50 in each group and decisive evidence when sample size is 100.

Table 2. Mean Bayes factors at each CpG site, based on simulation studies.

Start	End	N = 50 (Scenario 1)	N = 100 (Scenario 1)	N = 50 (Scenario 2)
1	6	1.02	1.02	1.03
2	7	1.01	1.02	1.01
3	8	1.01	1.02	1.02
4	9	1.02	1.01	1.01
5	10	1.24	1.53	1.26
6	11	1.78	3.12	1.78
7	12	2.95	9.16	2.85
8	13	5.74	41.42	4.95
9	14	10.53	1052.07	9.31
10	15	18.79	8554.12	18.31
11	16	13.9	3718.77	13.79
12	17	8.44	306.07	8.12
13	18	4.43	21.91	4.5
14	19	2.4	5.66	2.6
15	20	1.52	2.22	1.6
16	21	1.07	1.11	1.07
17	22	1.03	1.04	1.02
18	23	1.01	1.03	1.02
19	24	1.03	1.03	1.01

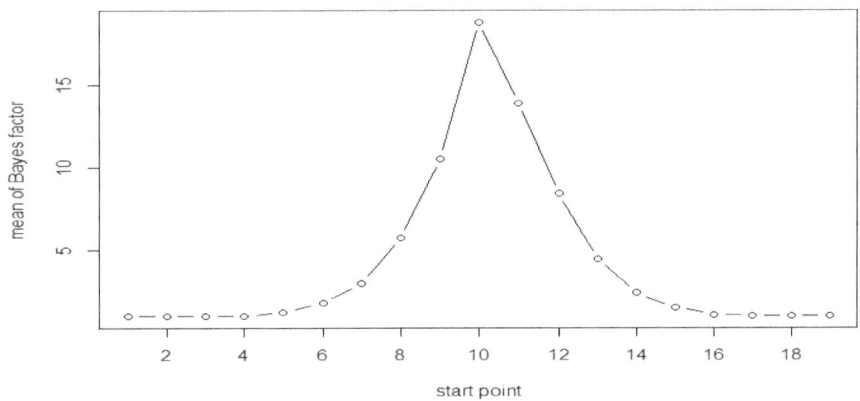

Figure 1. Mean of Bayes factors at each CpG site with $N = 50$ (Scenario 1).

All results show that the Bayes factors reach their maximum in the simulated DMR (CpG sites 10–15). However, the Bayes factors are not symmetric, the windows on the right side of the peak have larger values compared to those on the left side. This is attributed to the fact that the methylation status at a given site was generated conditional on that at the previous site of the same sequence. As expected, when the sample size is doubled the Bayes factors and the evidence in support of methylation increases significantly, as seen in Table 2 and Figures 1 and 2.

In order to illustrate that our proposed method is statistically valid and to ensure that the BF in our method is a meaningful measure for comparison with frequentist approaches, we computed Bayes factors exclusively for all moving windows that do not include the differentially methylated sites 10–11. Among these Bayes factors, 95% were less than 1.34 and 99% were less than 1.50, both consistent with Jeffreys' rule. These values can be thought of as the cut-offs corresponding to 5% and 1% empirical type I error rates. We calculated the proportions of times the Bayes factors fall above these cut-offs, for all possible numbers of DMCs in the moving window. These results are given in Table 3. For the simulated data they are comparable to the conclusions based on frequentist interpretations of type I error and power. For the real data analysis, one could employ a permutation test to derive the cutoff values under the null hypothesis. However, since the frequentist interpretation is not necessarily consistent with the Bayesian conclusions, using Jeffrey's rule for decision making may be more desirable when analyzing real data.

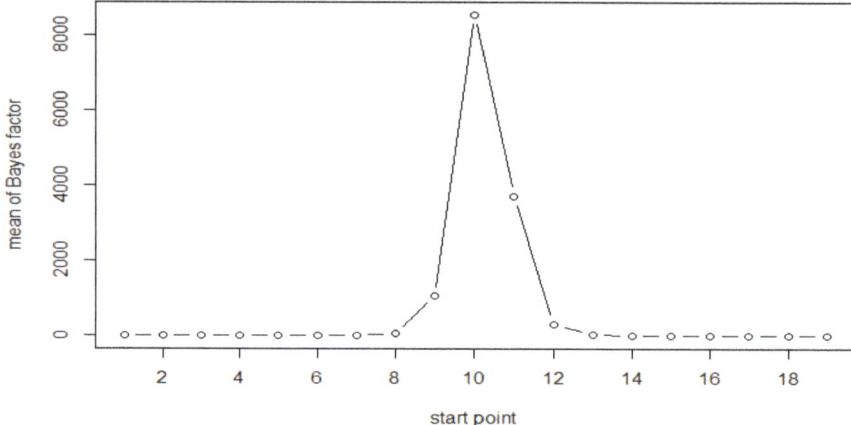

Figure 2. Mean of Bayes factors at each CpG site with $N = 100$ (Scenario 1).

Table 3. Proportions of Bayes factors that fell above the cut-off.

Cut-off Point	Number of DMCs in the Windows						
	0	1	2	3	4	5	6
1.34	0.050	0.56	0.97	1	1	1	1
1.5	0.010	0.35	0.91	1	1	1	1

3.2. Data analysis

We used our proposed BFM to analyze methylation data from a genome-wide association study of chronic lymphocytic leukemia (CLL), which manifests as a result of clonal expansion of malignant B cells. B-cell lymphoma, mostly prevalent among adults, is a heterogeneous disease [20,21]. It is clinically important to find heterogeneity of patients at the molecular level, which can help design specific interventions for patients at different severity levels.

Over the last decade, research in CLL has resulted in significant advances such as identification of several molecular alternations with prognostic values. These include specific cytogenetic patterns [22], mutational status of the immunoglobulin heavy chain variable gene (IgVH) [23] and expression of CD38 [24]. It has been found that patients lacking the mutation have a poorer prognosis. Patients with lower levels of CD38 have slower disease progression [23,25].

Several research groups have demonstrated that DNA methylation of multiple promoter-associated CpG islands is common in CLL [15,26,27]. Detection of aberrant DNA methylation in CLL could result in the development of an epigenetic classification of the disease with prognostic and therapeutic potential.

CD19+ B cells from peripheral blood were collected from CLL samples and normal control subjects. All CLL samples were obtained from patients at the Ellis Fischel Cancer Center (EFCC), the Georgia Cancer Center of Augusta University and the North Shore-LIJ Health System in compliance with the local Institutional Review Boards [28].

Illumina sequencing reads were generated for each sample by using RRBS [29]. In total, 20–30 million reads were sequenced for each sample, and 63%–75% were successfully mapped to either strand of the human genome (hg18) [28]. The average sequencing depth per CpG was between 32x and 43x. Eventually RRBS provided counts of DNA molecules that were methylated or unmethylated at each CpG site, and overall methylation status of approximately 1.8–2.3 million CpG sites were determined consistently for each sample in the study [28].

Tong et al. [30] pointed out that aberrant DNA methylation associated with CLL were located more frequently on chromosome 19. Hence, we analyzed genome-wide methylation data on 17,917 CpG sites on Chromosome 19 of 40 patients.

3.3. Comparison of Bayesian Method with Scan Statistic Method for Two Groups

First, we tested for differential methylation under binary response, by dividing the samples into two groups based on CD38 level of 20 as the cut-off. We had 23 subjects with CD38 $\leq$ 20 and 17 subjects with CD38 > 20. BFM and Scan statistic method (SSM) [31] were compared, using moving windows with 10 CpG sites in each window.

For comparing the two methods, we used a cut-off value of 2 for BFM and a 5% significance level for SSM. A total of 181 genes in DMRs were detected by SSM, and 183 genes were detected by BFM, using these criteria. Among these, 41 from SSM and 42 from BFM were found in PubMed publications as associated with leukemia (Table 4). There were 67 overlapping genes of which 18 were found in PubMed. They are ACP5, ATF5, BIRC8, C3, CARD8, CEACAM8, CERS1, CKM, CRTC1, IL4l1, LAIR1, MAP1S, NFIX, PDE4C, PLEKHG2, PLVAP, RFX1, and ZNF331 [32–49].

Table 4. Comparison of BFM and SSM for window size of 10 ($p < 0.05$).

	BFM > 2	SSM ($p < 0.05$)	Common
Total	183	181	67
PubMed	42	41	18

C3 and LAIR1((INK4a))genes were both detected, which were shown to be related to acute myeloid leukemia [34,41]. Actually, both C3 and LAIR1 genes connect with the transcription factor CREB (cyclic AMP response element binding protein), which has a role in the pathogenesis of AML and other cancers [50,51].

3.4. Bayesian Method for Ordinal Group Responses

To test whether the methylation rates increase as the CD38 levels increase, the samples were classified into four risk groups based on CD38 level, with 5 non-leukemia subjects in group 1, 23 patients in group 2 with CD38 $\leq$ 20, 9 patients in group 3 with 20 < CD38 $\leq$ 50, and 8 patients in group 4 with CD38 > 50. Though there are advantages of modeling CD38 as a continuous variable, but on the other hand, modeling as an ordinal variable is more robust to distributional assumptions. Again, moving windows of size of 10 were used for analysis. In fact, in clinical studies it is a common practice to put patients into discrete disease risk groups based on continuous measures.

Because of multiple testing issues associated with the comparison of four groups, we used a more stringent criterion of BF > 19 to evaluate the strength of evidence of differential methylation [8]. A total of 789 windows showed strong evidence of differential methylation using this criterion. The start and end positions in base pairs for each detected DMR were used in the UCSC genome browser to find the genes in the regions, and eventually 125 genes were found in these regions. Among them, 35 were associated with leukemia on PubMed literature. Some of these were not detected when only two groups were considered even with a less stringent criterion. They are BRD4, ELL, ERCC1, ERCC2, GDF15, JUND, POLD1, PRDX2, RANBP3, SPIB and TSPAN16 [52–62].

4. Discussion

Results from our simulation study indicate that BFM is a valid approach to detect DMRs when considering ordinal group responses, since the calculated Bayes factors were very large for simulated DMRs, and close to 1 for non-DMRs. The real data analysis based on the CLL data also demonstrated that BFM is a valid method that is able to detect DMRs with methylation rates increasing (or decreasing) as disease severity increases.

In addition to being able to account for ordering of group responses, BFM also has an advantage of allowing for heterogeneity of methylation effects across CpG sites by modeling the methylation rates with a prior. Methods such as the SSM pools information across variants in a region, assuming that each CpG sites in the region have the same methylation rates.

BFM with mixed-effect regression, not only can allow for covariates, but also the correlation between CpG sites. It takes advantage of the flexibility of the Bayesian framework, including the use of prior information when available as well as computational convenience, and uses distributions such as the multivariate normal to incorporate the correlation structure with inverse Wishart distribution as the prior for the correlation matrix.

One disadvantage of the BFM is that it assumes that methylation rates of CpG sites within each moving window are independent of those outside of the window, while the SSM accounts for the correlation along the whole genome.

BFM used a moving window to help decide the location and length of DMRs. But practically, it is very difficult to know the exact length of DMRs. This limitation is very common in statistical genetics, not only for detecting DMRs, but also for detecting rare variants [63]. Cross validation or bootstrap approaches might help determine the window sizes. It could be possible to develop other methods, for example, using genes and promoters instead of moving windows, along with the BFM to detect DMRs.

As described by George and Laud [64], the Bayes factor used in the context of testing hypotheses is a meaningful measure of evidence because it is a reasonably approximate factor by which the odds are increased by the data. With default priors that are essentially flat over a wide range of the relevant parameter space, the approach is similar to the likelihood-based inference. However, direct comparison between methods such as BFM based on the Bayes factor with frequentist approaches should be done with caution, as the Bayes factor classification for decision process is not a precise calibration, but rather a descriptive statement about the standards of evidence. Our proposed method is rather exploratory in nature, leading to a ranked list of sites for follow up for formal confirmation.

We developed the BFM, focusing only on DNA methylation data. However, large-scale cancer genomics projects such as TCGA (The Cancer Genome Atlas Research Network) are currently generating multiple layers of genomics data for early tumor, including DNA copy number, methylation, and mRNA expression. Similar statistical methods for integrated analysis and systematic modeling of these genomics data deserve further attention.

Author Contributions: Conceptualization, F.D., H.X. and V.G.; methodology, F.D., H.X, D.R., S.G. and V.G.; formal analysis, F.D.; writing—original draft preparation, F.D.; writing—H.X. and V.G.; data curation, H.S.

Funding: This research received no external funding.

Acknowledgments: The authors extend our special thanks and appreciation to one of the reviewers who provided a very extensive review of the manuscript with several insightful and constructive comments, which helped us improve the manuscript substantially.

Conflicts of Interest: The authors declare no conflict of interest.

References

1. Yokota, J. Tumor progression and metastasis. *Carcinogenesis* **2000**, *21*, 497–503. [CrossRef] [PubMed]
2. Torre, L.A.; Siegel, R.L.; Jemal, A. Lung Cancer Statistics. In *Lung Cancer and Personalized Medicine: Current Knowledge and Therapies*; Ahmad, A., Gadgeel, S., Eds.; Advances in Experimental Medicine and Biology; Springer International Publishing: Cham, Switzerland, 2016; pp. 1–19. ISBN 978-3-319-24223-1.

3. Qureshi, S.A.; Bashir, M.U.; Yaqinuddin, A. Utility of DNA methylation markers for diagnosing cancer. *Int J Surg* **2010**, *8*, 194–198. [CrossRef] [PubMed]
4. Varley, K.E.; Gertz, J.; Bowling, K.M.; Parker, S.L.; Reddy, T.E.; Pauli-Behn, F.; Cross, M.K.; Williams, B.A.; Stamatoyannopoulos, J.A.; Crawford, G.E.; et al. Dynamic DNA methylation across diverse human cell lines and tissues. *Genome Res.* **2013**, *23*, 555–567. [CrossRef] [PubMed]
5. Klajic, J.; Fleischer, T.; Dejeux, E.; Edvardsen, H.; Warnberg, F.; Bukholm, I.; Lønning, P.E.; Solvang, H.; Børresen-Dale, A.-L.; Tost, J.; et al. Quantitative DNA methylation analyses reveal stage dependent DNA methylation and association to clinico-pathological factors in breast tumors. *BMC Cancer* **2013**, *13*, 456. [CrossRef]
6. Watts, G.S.; Futscher, B.W.; Holtan, N.; Degeest, K.; Domann, F.E.; Rose, S.L. DNA methylation changes in ovarian cancer are cumulative with disease progression and identify tumor stage. *BMC Med Genomics* **2008**, *1*, 47. [CrossRef] [PubMed]
7. Hoque, M.O. DNA methylation changes in prostate cancer: current developments and future clinical implementation. *Expert Rev. Mol. Diagn.* **2009**, *9*, 243–257. [CrossRef] [PubMed]
8. Mitomi, H.; Fukui, N.; Tanaka, N.; Kanazawa, H.; Saito, T.; Matsuoka, T.; Yao, T. Aberrant p16 INK4a methylation is a frequent event in colorectal cancers: prognostic value and relation to mRNA expression and immunoreactivity. *J. Cancer Res. Clin. Oncol.* **2010**, *136*, 323–331. [CrossRef] [PubMed]
9. Dunson, D.B.; Neelon, B. Bayesian Inference on Order-Constrained Parameters in Generalized Linear Models. *Biometrics* **2003**, *59*, 286–295. [CrossRef]
10. Leek, J.T.; Scharpf, R.B.; Bravo, H.C.; Simcha, D.; Langmead, B.; Johnson, W.E.; Geman, D.; Baggerly, K.; Irizarry, R.A. Tackling the widespread and critical impact of batch effects in high-throughput data. *Nat. Rev. Genet.* **2010**, *11*, 733–739. [CrossRef]
11. Bartholomew, D.J. A test of homogeneity for ordered alternatives. *Biometrika* **1959**, *46*, 36–48. [CrossRef]
12. Robertson, T.; Wegman, E.J. Likelihood ratio tests for order restrictions in exponential families. *The Annals of Statistics* **1978**, 485–505. [CrossRef]
13. Ayer, M.; Brunk, H.D.; Ewing, G.M.; Reid, W.T.; Silverman, E. An empirical distribution function for sampling with incomplete information. *The annals of mathematical statistics* **1955**, 641–647. [CrossRef]
14. Taylor, J.M.G.; Wang, L.; Li, Z. Analysis on binary responses with ordered covariates and missing data. *Stat Med* **2007**, *26*, 3443–3458. [CrossRef] [PubMed]
15. Teschendorff, A.E.; Menon, U.; Gentry-Maharaj, A.; Ramus, S.J.; Weisenberger, D.J.; Shen, H.; Campan, M.; Noushmehr, H.; Bell, C.G.; Maxwell, A.P.; et al. Age-dependent DNA methylation of genes that are suppressed in stem cells is a hallmark of cancer. *Genome Res.* **2010**, *20*, 440–446. [CrossRef] [PubMed]
16. Kibriya, M.G.; Raza, M.; Jasmine, F.; Roy, S.; Paul-Brutus, R.; Rahaman, R.; Dodsworth, C.; Rakibuz-Zaman, M.; Kamal, M.; Ahsan, H. A genome-wide DNA methylation study in colorectal carcinoma. *BMC Med Genomics* **2011**, *4*, 50. [CrossRef] [PubMed]
17. Jeffreys, H. *Theory of probability, Clarendon*; Oxford University Press: Oxford, UK, 1961.
18. Kass, R.E.; Raftery, A.E. Bayes factors. *Journal of the american statistical association* **1995**, *90*, 773–795. [CrossRef]
19. Dai, B.; Ding, S.; Wahba, G. Multivariate bernoulli distribution. *Bernoulli* **2013**, *19*, 1465–1483. [CrossRef]
20. Chiorazzi, N.; Rai, K.R.; Ferrarini, M. Chronic lymphocytic leukemia. *N. Engl. J. Med.* **2005**, *352*, 804–815. [CrossRef]
21. Keating, M.J.; Chiorazzi, N.; Messmer, B.; Damle, R.N.; Allen, S.L.; Rai, K.R.; Ferrarini, M.; Kipps, T.J. Biology and treatment of chronic lymphocytic leukemia. *Hematology Am Soc Hematol Educ Program* **2003**, 153–175. [CrossRef]
22. Döhner, H.; Stilgenbauer, S.; Benner, A.; Leupolt, E.; Kröber, A.; Bullinger, L.; Döhner, K.; Bentz, M.; Lichter, P. Genomic aberrations and survival in chronic lymphocytic leukemia. *N. Engl. J. Med.* **2000**, *343*, 1910–1916. [CrossRef]
23. Hamblin, T.J.; Davis, Z.; Gardiner, A.; Oscier, D.G.; Stevenson, F.K. Unmutated Ig V(H) genes are associated with a more aggressive form of chronic lymphocytic leukemia. *Blood* **1999**, *94*, 1848–1854. [PubMed]
24. Hamblin, T.J.; Orchard, J.A.; Gardiner, A.; Oscier, D.G.; Davis, Z.; Stevenson, F.K. Immunoglobulin V genes and CD38 expression in CLL. *Blood* **2000**, *95*, 2455–2457. [PubMed]
25. Damle, R.N.; Wasil, T.; Fais, F.; Ghiotto, F.; Valetto, A.; Allen, S.L.; Buchbinder, A.; Budman, D.; Dittmar, K.; Kolitz, J.; et al. Ig V gene mutation status and CD38 expression as novel prognostic indicators in chronic lymphocytic leukemia. *Blood* **1999**, *94*, 1840–1847. [PubMed]

26. Kanduri, M.; Cahill, N.; Göransson, H.; Enström, C.; Ryan, F.; Isaksson, A.; Rosenquist, R. Differential genome-wide array-based methylation profiles in prognostic subsets of chronic lymphocytic leukemia. *Blood* **2010**, *115*, 296–305. [CrossRef] [PubMed]
27. Rahmatpanah, F.B.; Carstens, S.; Guo, J.; Sjahputera, O.; Taylor, K.H.; Duff, D.; Shi, H.; Davis, J.W.; Hooshmand, S.I.; Chitma-Matsiga, R.; et al. Differential DNA methylation patterns of small B-cell lymphoma subclasses with different clinical behavior. *Leukemia* **2006**, *20*, 1855–1862. [CrossRef]
28. Pei, L.; Choi, J.-H.; Liu, J.; Lee, E.-J.; McCarthy, B.; Wilson, J.M.; Speir, E.; Awan, F.; Tae, H.; Arthur, G.; et al. Genome-wide DNA methylation analysis reveals novel epigenetic changes in chronic lymphocytic leukemia. *Epigenetics* **2012**, *7*, 567–578. [CrossRef]
29. Meissner, A.; Gnirke, A.; Bell, G.W.; Ramsahoye, B.; Lander, E.S.; Jaenisch, R. Reduced representation bisulfite sequencing for comparative high-resolution DNA methylation analysis. *Nucleic Acids Res.* **2005**, *33*, 5868–5877. [CrossRef]
30. Tong, W.-G.; Wierda, W.G.; Lin, E.; Kuang, S.-Q.; Bekele, B.N.; Estrov, Z.; Wei, Y.; Yang, H.; Keating, M.J.; Garcia-Manero, G. Genome-wide DNA methylation profiling of chronic lymphocytic leukemia allows identification of epigenetically repressed molecular pathways with clinical impact. *Epigenetics* **2010**, *5*, 499–508. [CrossRef]
31. Dunbar, F.; Xu, H.; Ryu, D.; Ghosh, S.; Shi, H.; George, V. Computational Methods for Detection of Differentially Methylated Regions Using Kernel Distance and Scan Statistics. *Genes (Basel)* **2019**, *10*. [CrossRef]
32. French, D.; Hamilton, L.H.; Mattano, L.A.; Sather, H.N.; Devidas, M.; Nachman, J.B.; Relling, M.V. Children's Oncology Group A PAI-1 (SERPINE1) polymorphism predicts osteonecrosis in children with acute lymphoblastic leukemia: a report from the Children's Oncology Group. *Blood* **2008**, *111*, 4496–4499. [CrossRef]
33. Wang, T.; Qian, D.; Hu, M.; Li, L.; Zhang, L.; Chen, H.; Yang, R.; Wang, B. Human cytomegalovirus inhibits apoptosis by regulating the activating transcription factor 5 signaling pathway in human malignant glioma cells. *Oncol Lett* **2014**, *8*, 1051–1057. [CrossRef] [PubMed]
34. Glodkowska-Mrowka, E.; Solarska, I.; Mrowka, P.; Bajorek, K.; Niesiobedzka-Krezel, J.; Seferynska, I.; Borg, K.; Stoklosa, T. Differential expression of BIRC family genes in chronic myeloid leukaemia–BIRC3 and BIRC8 as potential new candidates to identify disease progression. *Br. J. Haematol.* **2014**, *164*, 740–742. [CrossRef] [PubMed]
35. Chae, H.-D.; Mitton, B.; Lacayo, N.J.; Sakamoto, K.M. Replication factor C3 is a CREB target gene that regulates cell cycle progression through the modulation of chromatin loading of PCNA. *Leukemia* **2015**, *29*, 1379–1389. [CrossRef] [PubMed]
36. Xu, W.; Zhou, L.; Chen, Q.; Chen, C.; Fang, L.; Fang, X.; Shen, H. [Effect of YB-1 gene knockdown on human leukemia cell line K562/A02]. *Zhonghua Yi Xue Yi Chuan Xue Za Zhi* **2009**, *26*, 400–405. [PubMed]
37. Lasa, A.; Serrano, E.; Carricondo, M.; Carnicer, M.J.; Brunet, S.; Badell, I.; Sierra, J.; Aventín, A.; Nomdedéu, J.F. High expression of CEACAM6 and CEACAM8 mRNA in acute lymphoblastic leukemias. *Ann. Hematol.* **2008**, *87*, 205–211. [CrossRef] [PubMed]
38. Camgoz, A.; Gencer, E.B.; Ural, A.U.; Baran, Y. Mechanisms responsible for nilotinib resistance in human chronic myeloid leukemia cells and reversal of resistance. *Leuk. Lymphoma* **2013**, *54*, 1279–1287. [CrossRef] [PubMed]
39. Caldow, M.K.; Digby, M.R.; Cameron-Smith, D. Short communication: Bovine-derived proteins activate STAT3 in human skeletal muscle in vitro. *J. Dairy Sci.* **2015**, *98*, 3016–3019. [CrossRef]
40. Tang, H.-M.V.; Gao, W.-W.; Chan, C.-P.; Cheng, Y.; Deng, J.-J.; Yuen, K.-S.; Iha, H.; Jin, D.-Y. SIRT1 Suppresses Human T-Cell Leukemia Virus Type 1 Transcription. *J. Virol.* **2015**, *89*, 8623–8631. [CrossRef] [PubMed]
41. Carbonnelle-Puscian, A.; Copie-Bergman, C.; Baia, M.; Martin-Garcia, N.; Allory, Y.; Haioun, C.; Crémades, A.; Abd-Alsamad, I.; Farcet, J.-P.; Gaulard, P.; et al. The novel immunosuppressive enzyme IL4I1 is expressed by neoplastic cells of several B-cell lymphomas and by tumor-associated macrophages. *Leukemia* **2009**, *23*, 952–960. [CrossRef]
42. Kang, X.; Lu, Z.; Cui, C.; Deng, M.; Fan, Y.; Dong, B.; Han, X.; Xie, F.; Tyner, J.W.; Coligan, J.E.; et al. The ITIM-containing receptor LAIR1 is essential for acute myeloid leukaemia development. *Nat. Cell Biol.* **2015**, *17*, 665–677. [CrossRef]
43. Haimovici, A.; Brigger, D.; Torbett, B.E.; Fey, M.F.; Tschan, M.P. Induction of the autophagy-associated gene MAP1S via PU.1 supports APL differentiation. *Leuk. Res.* **2014**, *38*, 1041–1047. [CrossRef] [PubMed]

44. O'Connor, C.; Campos, J.; Osinski, J.M.; Gronostajski, R.M.; Michie, A.M.; Keeshan, K. Nfix expression critically modulates early B lymphopoiesis and myelopoiesis. *PLoS ONE* **2015**, *10*, e0120102. [CrossRef] [PubMed]
45. Moon, E.; Lee, R.; Near, R.; Weintraub, L.; Wolda, S.; Lerner, A. Inhibition of PDE3B augments PDE4 inhibitor-induced apoptosis in a subset of patients with chronic lymphocytic leukemia. *Clin. Cancer Res.* **2002**, *8*, 589–595. [PubMed]
46. Runne, C.; Chen, S. PLEKHG2 promotes heterotrimeric G protein βγ-stimulated lymphocyte migration via Rac and Cdc42 activation and actin polymerization. *Mol. Cell. Biol.* **2013**, *33*, 4294–4307. [CrossRef] [PubMed]
47. Rantakari, P.; Auvinen, K.; Jäppinen, N.; Kapraali, M.; Valtonen, J.; Karikoski, M.; Gerke, H.; Iftakhar-E-Khuda, I.; Keuschnigg, J.; Umemoto, E.; et al. The endothelial protein PLVAP in lymphatics controls the entry of lymphocytes and antigens into lymph nodes. *Nat. Immunol.* **2015**, *16*, 386–396. [CrossRef] [PubMed]
48. Chen, L.; Smith, L.; Johnson, M.R.; Wang, K.; Diasio, R.B.; Smith, J.B. Activation of protein kinase C induces nuclear translocation of RFX1 and down-regulates c-myc via an intron 1 X box in undifferentiated leukemia HL-60 cells. *J. Biol. Chem.* **2000**, *275*, 32227–32233. [CrossRef] [PubMed]
49. McHale, C.M.; Zhang, L.; Lan, Q.; Li, G.; Hubbard, A.E.; Forrest, M.S.; Vermeulen, R.; Chen, J.; Shen, M.; Rappaport, S.M.; et al. Changes in the peripheral blood transcriptome associated with occupational benzene exposure identified by cross-comparison on two microarray platforms. *Genomics* **2009**, *93*, 343–349. [CrossRef]
50. Crans-Vargas, H.N.; Landaw, E.M.; Bhatia, S.; Sandusky, G.; Moore, T.B.; Sakamoto, K.M. Expression of cyclic adenosine monophosphate response-element binding protein in acute leukemia. *Blood* **2002**, *99*, 2617–2619. [CrossRef]
51. Mayr, B.; Montminy, M. Transcriptional regulation by the phosphorylation-dependent factor CREB. *Nat. Rev. Mol. Cell Biol.* **2001**, *2*, 599–609. [CrossRef]
52. Stewart, H.J.S.; Horne, G.A.; Bastow, S.; Chevassut, T.J.T. BRD4 associates with p53 in DNMT3A-mutated leukemia cells and is implicated in apoptosis by the bromodomain inhibitor JQ1. *Cancer Med* **2013**, *2*, 826–835. [CrossRef]
53. Muto, T.; Takeuchi, M.; Yamazaki, A.; Sugita, Y.; Tsukamoto, S.; Sakai, S.; Takeda, Y.; Mimura, N.; Ohwada, C.; Sakaida, E.; et al. Efficacy of myeloablative allogeneic hematopoietic stem cell transplantation in adult patients with MLL-ELL-positive acute myeloid leukemia. *Int. J. Hematol.* **2015**, *102*, 86–92. [CrossRef] [PubMed]
54. Kong, J.H.; Mun, Y.-C.; Kim, S.; Choi, H.S.; Kim, Y.-K.; Kim, H.-J.; Moon, J.H.; Sohn, S.K.; Kim, S.-H.; Jung, C.W.; et al. Polymorphisms of ERCC1 genotype associated with response to imatinib therapy in chronic phase chronic myeloid leukemia. *Int. J. Hematol.* **2012**, *96*, 327–333. [CrossRef] [PubMed]
55. Liu, D.; Wu, D.; Li, H.; Dong, M. The effect of XPD/ERCC2 Lys751Gln polymorphism on acute leukemia risk: a systematic review and meta-analysis. *Gene* **2014**, *538*, 209–216. [CrossRef] [PubMed]
56. Secchiero, P.; Barbarotto, E.; Tiribelli, M.; Zerbinati, C.; di Iasio, M.G.; Gonelli, A.; Cavazzini, F.; Campioni, D.; Fanin, R.; Cuneo, A.; et al. Functional integrity of the p53-mediated apoptotic pathway induced by the nongenotoxic agent nutlin-3 in B-cell chronic lymphocytic leukemia (B-CLL). *Blood* **2006**, *107*, 4122–4129. [CrossRef] [PubMed]
57. Gazon, H.; Lemasson, I.; Polakowski, N.; Césaire, R.; Matsuoka, M.; Barbeau, B.; Mesnard, J.-M.; Peloponese, J.-M. Human T-cell leukemia virus type 1 (HTLV-1) bZIP factor requires cellular transcription factor JunD to upregulate HTLV-1 antisense transcription from the 3′ long terminal repeat. *J. Virol.* **2012**, *86*, 9070–9078. [CrossRef] [PubMed]
58. Sincennes, M.-C.; Humbert, M.; Grondin, B.; Lisi, V.; Veiga, D.F.T.; Haman, A.; Cazaux, C.; Mashtalir, N.; Affar, E.B.; Verreault, A.; et al. The LMO2 oncogene regulates DNA replication in hematopoietic cells. *Proc. Natl. Acad. Sci. U.S.A.* **2016**, *113*, 1393–1398. [CrossRef]
59. Agrawal-Singh, S.; Isken, F.; Agelopoulos, K.; Klein, H.-U.; Thoennissen, N.H.; Koehler, G.; Hascher, A.; Bäumer, N.; Berdel, W.E.; Thiede, C.; et al. Genome-wide analysis of histone H3 acetylation patterns in AML identifies PRDX2 as an epigenetically silenced tumor suppressor gene. *Blood* **2012**, *119*, 2346–2357. [CrossRef]
60. Hakata, Y.; Yamada, M.; Shida, H. A multifunctional domain in human CRM1 (exportin 1) mediates RanBP3 binding and multimerization of human T-cell leukemia virus type 1 Rex protein. *Mol. Cell. Biol.* **2003**, *23*, 8751–8761. [CrossRef]

61. Talby, L.; Chambost, H.; Roubaud, M.-C.; N'Guyen, C.; Milili, M.; Loriod, B.; Fossat, C.; Picard, C.; Gabert, J.; Chiappetta, P.; et al. The chemosensitivity to therapy of childhood early B acute lymphoblastic leukemia could be determined by the combined expression of CD34, SPI-B and BCR genes. *Leuk. Res.* **2006**, *30*, 665–676. [CrossRef]
62. Juric, D.; Lacayo, N.J.; Ramsey, M.C.; Racevskis, J.; Wiernik, P.H.; Rowe, J.M.; Goldstone, A.H.; O'Dwyer, P.J.; Paietta, E.; Sikic, B.I. Differential gene expression patterns and interaction networks in BCR-ABL-positive and -negative adult acute lymphoblastic leukemias. *J. Clin. Oncol.* **2007**, *25*, 1341–1349. [CrossRef]
63. Schaid, D.J.; Sinnwell, J.P.; McDonnell, S.K.; Thibodeau, S.N. Detecting genomic clustering of risk variants from sequence data: cases versus controls. *Hum. Genet.* **2013**, *132*, 1301–1309. [CrossRef] [PubMed]
64. George, V.; Laud, P.W. A Bayesian approach to the transmission/disequilibrium test for binary traits. *Genetic Epidemiol.* **2002**, *22*, 41–51. [CrossRef] [PubMed]

 © 2019 by the authors. Licensee MDPI, Basel, Switzerland. This article is an open access article distributed under the terms and conditions of the Creative Commons Attribution (CC BY) license (http://creativecommons.org/licenses/by/4.0/).

Article

Local Epigenomic Data are more Informative than Local Genome Sequence Data in Predicting Enhancer-Promoter Interactions Using Neural Networks

Mengli Xiao [1], Zhong Zhuang [2] and Wei Pan [1,*]

[1] Division of Biostatistics, University of Minnesota, Minneapolis, MN 55455, USA; xiaox345@umn.edu
[2] Department of Electrical and Computer Engineering, University of Minnesota, Minneapolis, MN 55455, USA; zhuan143@umn.edu
* Correspondence: panxx014@umn.edu; Tel.: +01-612-626-2705

Received: 29 November 2019; Accepted: 26 December 2019; Published: 29 December 2019

Abstract: Enhancer-promoter interactions (EPIs) are crucial for transcriptional regulation. Mapping such interactions proves useful for understanding disease regulations and discovering risk genes in genome-wide association studies. Some previous studies showed that machine learning methods, as computational alternatives to costly experimental approaches, performed well in predicting EPIs from local sequence and/or local epigenomic data. In particular, deep learning methods were demonstrated to outperform traditional machine learning methods, and using DNA sequence data alone could perform either better than or almost as well as only utilizing epigenomic data. However, most, if not all, of these previous studies were based on randomly splitting enhancer-promoter pairs as training, tuning, and test data, which has recently been pointed out to be problematic; due to multiple and duplicating/overlapping enhancers (and promoters) in enhancer-promoter pairs in EPI data, such random splitting does not lead to independent training, tuning, and test data, thus resulting in model over-fitting and over-estimating predictive performance. Here, after correcting this design issue, we extensively studied the performance of various deep learning models with local sequence and epigenomic data around enhancer-promoter pairs. Our results confirmed much lower performance using either sequence or epigenomic data alone, or both, than reported previously. We also demonstrated that local epigenomic features were more informative than local sequence data. Our results were based on an extensive exploration of many convolutional neural network (CNN) and feed-forward neural network (FNN) structures, and of gradient boosting as a representative of traditional machine learning.

Keywords: boosting; convolutional neural networks; deep learning; feed-forward neural networks; machine learning

1. Introduction

Non-coding genome sequences, including enhancers, promoters, and other regulatory elements, play important roles in transcriptional regulation. In particular, through enhancer-promoter interactions (i.e., physical contacts), the enhancers and promoters coordinately regulate gene expression. Although enhancers can be distal from promoters in the genome, they are brought close to, and possibly in contact with, each other in the 3-D space through chromatin looping. Some enhancers even bypass adjacent promoters to interact with the target promoters in response to histone or transcriptional modifications on the genome. An accurate mapping of such distant interactions is of particular interest for understanding gene expression pathways and identifying target genes of GWAS loci [1–3].

Experimental methods based on chromosome conformation capture (3C, 4C, and Hi-C) or extensions that incorporate ChIP-sequencing such as paired-end tag sequencing (ChIA-PET) are, however, costly, and the results are only available for a few cell types [4–7]. Computational tools offer an alternative by utilizing various DNA sequence and/or epigenomic annotation data to predict EPIs with machine learning models built from experimentally obtained EPI data [8–11].

Whalen, et al. [11] reported that a gradient boosting method, called TargetFinder, accurately distinguished between interacting and non-interacting enhancer-promoter pairs based on epigenomic profiles. They included histone modifications and transcription factor binding (based on ChIP-seq), and DNase I hypersensitive sites (DNase-seq) with a focus on distal interaction (>10 kb) in high resolution. The idea was further extended to predict EPIs solely from local DNA sequence data and achieved high prediction accuracy [12–14].In particular, convolutional neural networks (CNNs), known for capturing stationary patterns in data with successful applications in image and text recognition [15,16], were shown to perform well in predicting EPIs based on DNA sequence alone. A natural question is whether CNNs can further improve the prediction performance with regional epigenomic features. It is also noted that for DNA sequence data, differing from for images, a simple CNN model seemed to perform well [14]; a similar conclusion was drawn for other biological data [17].

On the other hand, two recent studies pointed out an experimental design issue of randomly splitting the original data for training and testing as adopted by most, if not all, previous studies: many promoters interact with multiple, possibly overlapping, enhancers concurrently. Such pairs, some in the training data while others in the test data, are not independent, leading to possibly over-fitting a model and over-estimating its predictive performance [18,19] Since promoters primarily interact with enhancers on the same chromosome, the problem could be avoided by having different chromosomes split into the training and test data. Based on such a valid training and test data, Xi and Beer [19] concluded that local epigenomic features around enhancers and promoters alone were not informative enough to predict EPIs and they suggested to re-evaluate similar studies on EPI prediction. Although combining local epigenomic features and sequence data was found to improve prediction in a recent study [20] it was based on a random splitting of the whole dataset into training and test data, thus possibly suffering from inflated performance. After correcting for such experimental design bias, we would like to address two important and interesting questions: (1) whether or not local enhancer-promoter sequences are more informative than corresponding local epigenomic features; (2) whether or not we can gain more information by combining local sequence and epigenomic annotations.

Here, we report our extensive study of local sequence and epigenomic data for predicting long-range EPIs; in addition to more recent and popular CNNs and gradient boosting as adopted in most of prior studies, we also considered more traditional feed-forward neural networks (FNNs) [21,22]. After avoiding the previous experimental design issue, we found that local sequence data alone were insufficient to predict EPIs well; in comparison, local epigenomic signals, albeit not highly predictive either, were more informative than sequence data. Furthermore, combining local sequence data with local epigenomic profiles did not improve over using local epigenomic data alone. These results may be useful for future studies.

2. Materials and Methods

In this paper, we primarily studied one of the six cell lines, K562, with a large sample size and top performance in previous studies among the six cell lines, to investigate within-cell-line predictive performance [11]. The sample size was 41,477 in total with 1977 positive cases (i.e., interactions) and 39,500 negative cases.

2.1. Data

We retrieved the DNA sequence data packaged in the SPEID model of Singh, et al. [13] at the website http://genome.compbio.cs.cmu.edu/~{}sss1/SPEID/all_sequence_data.h5. Each sample represented an interacting/non-interacting pair of one-hot encoded DNA sequence centered at one enhancer (3000 bp) and one promoter (2000 bp), which were only local features compared to >10 kb distance between an enhancer and a promoter in the dataset [11] As for the local sequence data used previously [13,14], epigenomic data length was set to be 3000 bp for enhancers and 2000 bp for promoters in all epigenomic data types. We used the epigenomic features that are shared across the cell lines with most genomic features (K562, GM12878, HeLa-S3, and IMR90) based on Supplemental Table 2 in Whalen, et al. (2016). There are 22 epigenomic data types in total, including 11 histone mark peaks (H3K27ac,H3K27me3,H3K4me1, H2AZ, H3K4me2, H3K9ac, H3K4me3, H4K20me1, H3K79me2, H3K36me3, H3K9me3), 9 transcriptional factor bindings (POLR2A, CTCF, EP300, MAFK, MAZ, MXI1, RAD21, RCOR1, RFX5), DNase and methylation (ENCODE Project Consortium, 2007; Bernstein, 2010). Hence, the data dimensions were (# of samples)×3000×22 for enhancers and (# of samples×2000×22) for promoters (Figures 1–3). We extracted the epigenomic data as following. According to the enhancer/promoter genome coordinates available in the TargetFinder E/P dataset, the 3000 and 2000 genome window coordinates centered around the enhancer and promoter in each pair were calculated, then the data at each base pair were retrieved via those calculated window coordinates from 22 epigenomic data files across the whole genome for cell line K562 in the BigWig format, available at the ENCODE or NIH Roadmap Epigenomic projects [23,24]. We used package pyBigWig (http://dx.doi.org/10.5281/zenodo.45238) to read in BigWig files. However, one type of epigenomic data file in the BigWig format was often measured with multiple sample replicates, but the ENCODE or Roadmap project summarized those measurements only in BED format. The two file formats contain epigenomic feature information in different genome scales. Each unit in the BED file represents a small sampled genome sub-region with experimentally measured signals, thus the same base pair may be measured multiple times in multiple and different sub-region samples, which can be combined to map a unique signal value to each base pair by available tools for the whole genome. BigWig file, however, has a 1-1 correspondence between one base pair and a signal value across the whole genome. Thus, we obtained such 1-1 map between signals and genome in BigWig format data from BED files in Whalen, et al. [11] at https://github.com/shwhalen/targetfinder/blob/master/paper/targetfinder/K562/, which came from the cleaned peak files through the ENCODE or Roadmap. The Bedtools and bedGraphToBigWig software (http://hgdownload.soe.ucsc.edu/admin/exe/linux.x86_64/) was used to merge and convert BED files to the BigWig format [25]. To compare with TargetFinder, where data was summarized by computing the mean signal value across the whole local region of interest (3000/2000 bp), we also took the mean of the local epigenomic signals across each 3000/2000-bp window Later we will refer it as the TargetFinder-format data, which was 2-dimensional (Figure 4), and not suitable for CNNs. The epigenomic data were large to read in (13–23 GB) and often sparse in a 3000/2000-bp interval. Even when not sparse, the signal values often remained the same across multiple base pairs. The redundant and noisy data unnecessarily increased the number of parameters in prediction models. Therefore, we considered averaging the data signals through different sliding-window sizes and step sizes. Through the validation performance of CNNs, we found that a window/bin size of 50 and step size of 10 performed well among non-summarized and other forms of summarized data (Table S1). The data were later input into CNNs, hence we call it CNN-format data (Figure 2). After the sliding-window operation, the CNN-format data were reduced to 1–2 GB.

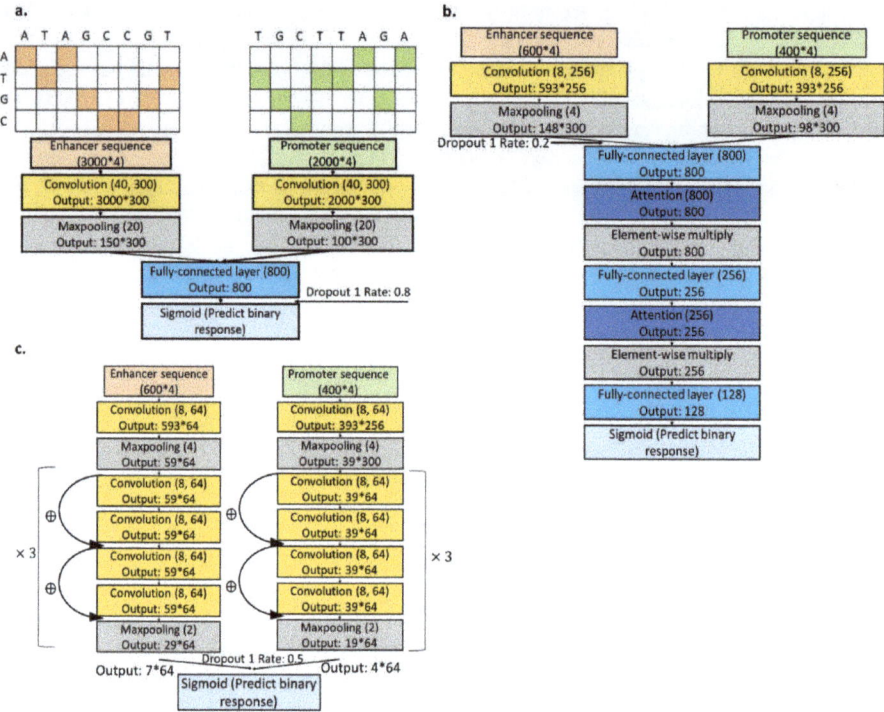

Figure 1. Sequence model structures. (**a**) The structure followes a previously reported simple convolutional model [14]. We used shorter input sequences, each centered as the SPEID data, with attention modules (**b**) or ResNet (**c**).

To combine sequence and epigenomic data sources, a one-to-one mapping from enhancer-promoter pairs in the TargetFinder dataset to SPEID sequences was established. Although the exact procedure in the Singh, et al. [13] about sequence data generation from the TargetFinder E/P dataset was not available online, we inferred the relationship by matching sequences. Given the provided enhancer/promoter locations on the genome in TargetFinder, we first retrieved the enhancer and promoter sequence segments from the hg19 reference genome at http://genome.ucsc.edu/cgi-bin/das/hg19/dna?segment=chr1:6454864,6455189 (e.g., an enhancer was located on chromosome 1 from position 6454864 to 6455189 bp), then searched a matching SPEID enhancer/promoter sequence.

We always used the data on chromosomes 8 and 9 as the validation data; we used each of the remaining chromosome in turn as a test dataset while the other chromosomes (not the three chromosomes used as the validation and test data) as the training data. This avoided the bias issue of the original random data splitting [19].

2.2. Designing Neural Networks for Utilizing Enhancer and Promoter Features

Our main goal was to construct predictive CNN models for detecting distal EPIs from sequence and epigenomic data, separately or combined. CNNs for sequence and epigenomic annotations were built separately, and we constructed another model to combine features from the two sources. In either sequence or epigenomics CNN, we considered two separate branches for enhancers and promoters respectively, then concatenated the features later. This followed from that enhancers and promoters are expected to have different sequence motifs or epigenomic profiles, as shown in previously designed CNNs [13,14,20].

2.3. Sequence CNNs.

Figure 1 shows different CNN models for sequence data after splitting the data by chromosomes to prevent non-independent training, validation, and test data. We conducted an extensive evaluation of CNNs with varying structures and parameters (Table 1 and Table S2). We first tested the performance of the CNN in Zhuang, et al. [14], and then added an attention module to utilize the sequence structure and focus only on the middle part of the input sequence [26]. Moreover, we tested on the use of the Residual Neural Network (ResNet) architecture for the data (Table 2) [27,28].

Table 1. Parameters for convolutional neural network (CNN) models.

	Batch Size	Learning Rate	L2 Weight Decay in the Convolution	Dropout 1	Dropout 2
		Sequence			
Zhuang et al., 2019 [14] (Basic CNN)	64	1.0×10^{-5}	1.0×10^{-6}	0.2	0.8
Attention CNN	64	1.0×10^{-5}	2.0×10^{-5}	0.2	None
ResNet CNN	200	1.0×10^{-5}	2.0×10^{-5}	0.5	None
		Epigenomics			
Basic CNN	200	1.0×10^{-6}	1.0×10^{-5}	0.3	0.6
ResNet CNN	64	1.0×10^{-3}	1.0×10^{-5}	0.3	0.6
FNN	200	1.0×10^{-6}	1.0×10^{-5}	0.7	None
Combined model	200	1.0×10^{-5}	1.0×10^{-6}	0.3	0.5

2.4. Epigenomics CNNs

A 4-layer CNN (called basic CNN) and a 10-layer ResNet are shown in Figure 2, chosen by the validation performance to predict EPIs based on epigenomic data (Table 3 and Table S2). We also implemented a few other newer CNN models, including Inception and Capsule Networks because of their improved performance in image applications [29,30]. Likely due to limited epigenomic data, their results were less competitive than simpler CNNs; they provided no higher prediction power than random guessing (result not shown).

Table 2. Sequence model structures.

Block Name	Simple/Basic CNN	Attention	ResNet without Fully-Connected Layer
Input	Enhancer : 3000×4 Promoter : 2000×4	Enhancer : 600×4 Promoter : 400×4	Enhancer : 600×4 Promoter : 400×4
conv1_x	$40 \times 1, 300$ maxpooling(20×1)	$8 \times 1, 256$ maxpooling(4×1)	$8 \times 1, 64$ maxpooling(10×1)
conv2_x			$\begin{bmatrix} 8 \times 1, 64 \\ 8 \times 1, 64 \end{bmatrix} \times 2$
conv3_x			$\begin{bmatrix} 8 \times 1, 64 \\ 8 \times 1, 64 \end{bmatrix} \times 2$
conv4_x			$\begin{bmatrix} 8 \times 1, 64 \\ 8 \times 1, 64 \end{bmatrix} \times 2$
Concatenate branches	800-d fc, sigmoid	800 − dfc 800 − dattention multiplytheoutputs 256 − dfc 256 − dattention multiplytheoutputs 128 − dfc Sigmoid	noavgpooling(outputdimesnionis11) Sigmoid
# of parameters	60,100,402	51,345,538	605,452

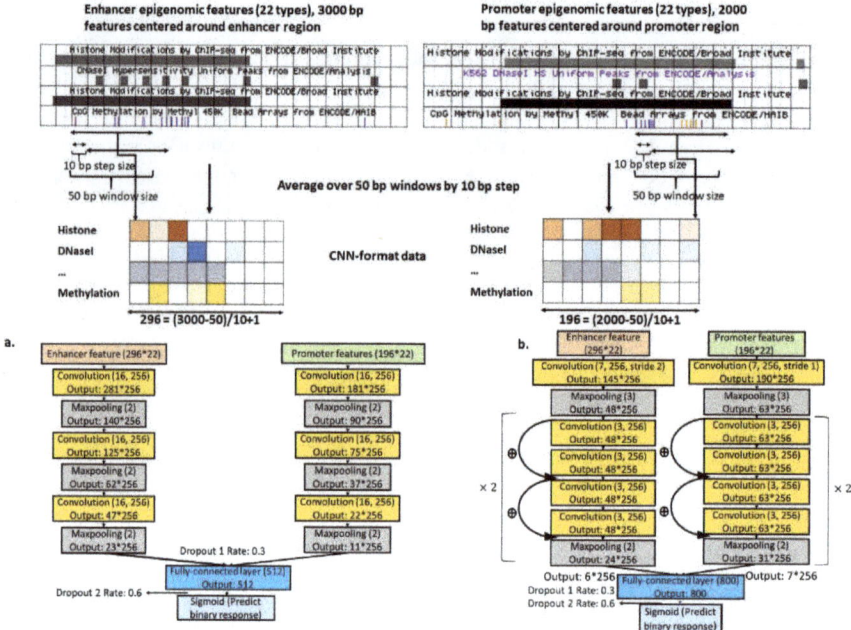

Figure 2. Epigenomics data preparation and CNN model structures. Different genomic annotation data type was denoted with different color. (**a**) Basic CNN structure; (**b**) ResNet CNN structure.

2.5. Combined Models

To combine the features from two data sources, we constructed 1 or 2 hidden layers (Figure 3 and Table S2). We saved high-level features extracted from the previous CNNs for sequence and epigenomic data, then combined them together as input to the first hidden layer in the combined model. We also tested more complex combination schemes but found no performance improvement; for example, we trained and combined the individual models simultaneously, instead of doing the two steps sequentially.

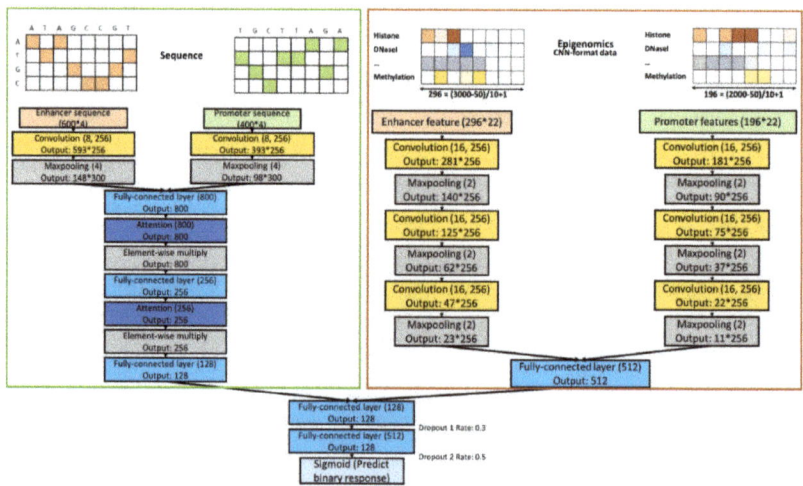

Figure 3. Combined model structure with the sequence and epigenomic models.

2.6. Designing FNNs.

Since an FNN can take epigenomic data as input in the same format as that of TargetFinder (i.e., TargetFinder-format data), we were interested to know if such a simple FNN could outperform TargetFinder or other CNNs. After tuning the model parameters through the validation dataset, we chose a 2-layer FNN shown in Figure 4.

An FNN can take both TargetFinder-format data and CNN-format data as input, and our analysis showed an FNN with the same structure performed better with the TargetFinder-format data than that with the CNN-format data (Table S3). Therefore, we further explored the performance of basic CNNs shown in Figure 2 and FNNs shown in Figure 4 in comparison to gradient boosting as implemented in TargetFinder [11]. We not only implemented neural networks for K562 cell line, but also expanded our evaluations to other cell lines (GM12878, HeLa-S3, IMR90). The training configuration was similar to that for CNNs.

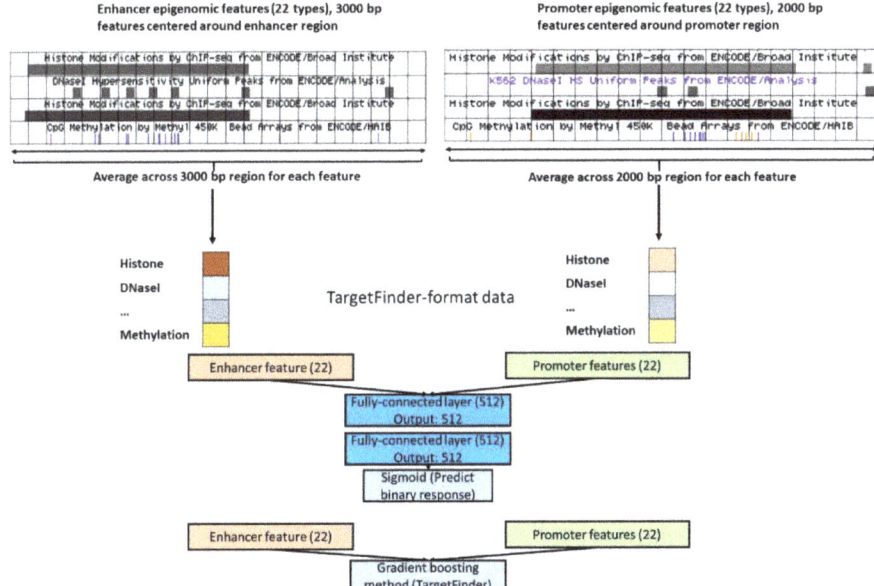

Figure 4. TargetFinder-format (epigenomic) data and the feed-forward neural network (FNN) model structure.

2.7. CNN Training in an Imbalanced-Class Scenario

For each CNN model in our study, the number of model parameters was much larger than the available training sample size, which might cause overfitting and numerical instability. In addition, there was a high ratio of class imbalance (positive:negative = 1:20), posing another challenge to training predictive models. Different training techniques were adopted to address these problems. Batch normalization was added after each layer to stabilize gradient updates and reduce the dependence on initialization [31]. The neural network weights were initialized from the glorot uniform distribution: U(-sqrt(6/(# of input weights+# of output weights)), sqrt(6/(# of input weights+# of output weights))) or He's normal distribution in ResNet for all layers [32,33]. The parameters were estimated through the Adam optimizer [34]. The learning rate and batch size were tuned across multiple grid values (Table S2). We used the weight decay and dropout for regularization [35], and more details about our final training parameters for all main CNNs are in Table 1. Another regularization technique is the early stopping that was determined by the validation data on chromosomes 8 and 9; the training was stopped if no improvement of the validation F1 score was observed over 10 epochs. Final evaluation on the test data used the model with the highest validation F1 score across all the training epochs before early stopping.

In order to address the problem of highly imbalanced classes in the training data, we used a weighted objective function—a binary cross entropy for this imbalanced dataset. In the training data, the weight for positive or negative pairs was given by the ratio of a half of the sample size over the number of positive or negative pairs. We also attempted to achieve a balanced training sample by augmenting the training data through oversampling positive pairs (i.e., the minority class) or down-sampling negative pairs in generating batched training data for neural networks, but the test performance was not significantly better than using a weighted objective function (with a p value of 0.4097 from the paired t-test; Table S4).

2.8. Evaluating Model Performance

With local sequence and 22 types of epigenomic features in our highly imbalanced dataset, the prediction performance was assessed through an Area Under Receiver Operating Characteristic (AUROC) with test data. Note that we had 21 test datasets corresponding to 21 chromosomes besides chromosomes 8 and 9, which were used as the validation data. Since the numbers of enhancer-promoter pairs varied with chromosomes, we reported a weighted average and standard deviation of the AUROC's across 21 chromosomes, where the weight was the (# of samples on each chromosome)/(total # of samples on 21 chromosomes).

To examine if either sequence or epigenomics model prediction performance could benefit from the other data source, we also evaluated the chromosome-wise test AUROCs for the combined model structure in Figure 4. Data splitting and training configurations (including tuning both structural and training parameters) followed exactly as before. We combined the models with the highest mean AUROCs from the two data sources respectively in Table 4, which were the sequence attention model (central region) and epigenomics basic CNN model. We also conducted the paired t-test across the 21 test chromosomes for the difference between two methods (with a null hypothesis H_0 that two methods gave the same test AUROC for each chromosome). The p values reported here were not adjusted for multiple comparisons, but would remain significant (or insignificant) after the Bonferroni adjustment.

2.9. Implementation and Code Availability

All the neural networks below were implemented in Keras (2.0.9) with Tensorflow (1.4.0) on a GPU server (NVIDIA Telsa K40 GPU). The choice of the parameters for gradient boosting followed TargetFinder (https://github.com/shwhalen/targetfinder), and we chose the # of trees through the validation dataset. Gradient boosting was implemented with Python Sklearn. The code is available at https://github.com/menglix/EPI.

3. Results

3.1. Local Epigenomic Features Were more Informative than Local Sequence Data in Predicting EPIs

Although previous studies found using sequence data yielded as good or even better prediction performance as/than using epigenomic data [13,14], this trend was reversed after a valid data splitting scheme was applied (Methods). Figure 5a shows that using (local) epigenomic data outperformed using (local) sequence data across all test chromosomes for each of multiple prediction models. We also compared the performance of the two data sources with similar models side by side in Figure 5b, where the basic CNN, ResNet CNN and gradient boosting were customized to the two data sources during the training process. p-values of the paired t-test (for each chromosome) to compare model weighted average AUROC were all <0.0002, suggesting that the local epigenomics data gave a statistically significant and stronger performance than the local sequence data.

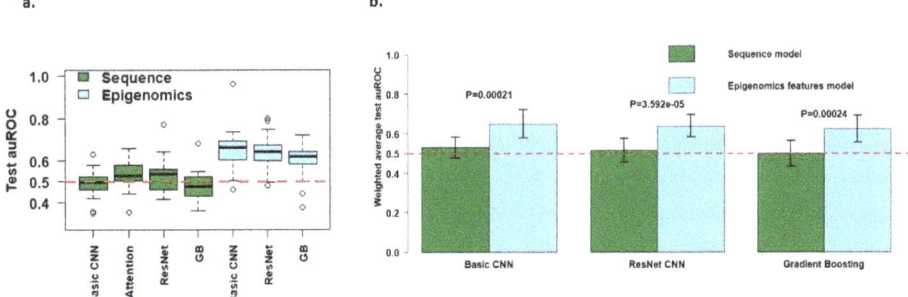

Figure 5. Comparison on the weighted average Area Under Receiver Operating Characteristic (AUROC) between the sequence and epigenomics models. (**a**) Boxplot of test AUROCs for all sequence and epigenomic models across 21 chromosomes; (**b**) Mean test AUROC comparison between the sequence and epigenomics models. The bars are ±1 weighted standard deviation around the weighted mean AUROC. p values are from the paired t-test with H_0: The test AUROCs are the same for the sequence and epigenomics models. p values are not adjusted for multiple comparison but remained significant after the Bonferroni adjustment.

Interestingly, while the test AUROC's for the sequence models were very close to 0.5, corresponding to random guessing, the epigenomic models achieved a notable difference from 0.5 (Figure 5a,b). This again suggests a better performance of using the epigenomics data among all prediction models implemented here. Furthermore, the AUROC's of the epigenomics models had similar standard deviations across chromosomes to those of the sequence models, and even minus one standard deviation of the mean AUROC of the epigenomics models was above the sequence models' average AUROC for each of the 3 scenarios in Figure 5b. Since the epigenomics models' AUROCs were significantly different from that of the sequence models, which were comparable to random guessing (0.5), local epigenomics features were still predictive of EPIs, though the performance was much inflated in previous studies [11,19,20]. Our finding also demonstrated that without appropriate data splitting (or generally valid experimental design), any EPI prediction results and downstream motif analyses with DNA sequence data should be interpreted with caution.

3.2. Combining Epigenomic and Sequence Data: Do We Gain Additional Information?

As shown in Table 4 and Figure 6, the combined model showed an improved mean AUROC over that of the sequence model (AUROC of 0.603 vs. 0.529), though the result was expected as epigenomics features showed a higher predictive power than sequences from Figure 5. Still, our finding was consistent with other publications showing integrating epigenomic features with sequence data improved the performance over using sequence data alone in predicting epigenomics-related features with fully-connected layers or recurrent neural network [20,36,37]. However, the performance was not enhanced over that of the epigenomics model (AUROC of 0.603 vs. 0.648).

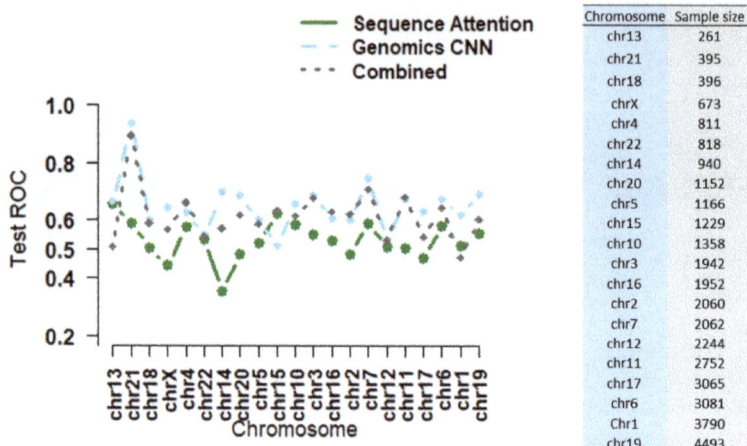

Figure 6. The combined CNN model performance as compared to the sequence and epigenomics CNNs across 21 test chromosomes with the sample size in ascending order.

3.3. Are more Complex Structures and more Parameters Needed for High-Dimensional Data Input?

Our input data for CNN models were relatively high-dimensional (with the enhancer sequence and epigenomic features of dimensions of 3000 × 4 and 296 × 22 respectively). Through our extensive explorations of various CNN architectures/structures, we observed that both epigenomic data- and sequence data-based prediction models performed better with simpler CNNs (Tables 2 and 3). At the same time, we still needed a large number of parameters in each layer, leading to over-parametrized models. For example, in the epigenomic ResNet model, although the highest validation AUROC required only 2 blocks (i.e., 8 convolution layers in Table 3), it requires a large number of filters (256) according to Figure 7. Several lines of evidence during our model tuning supported our conclusions. First, our observed optimized numbers of layers in CNNs were small, in contrast to deep learning models in image recognition and other applications (Tables 2 and 3). This was perhaps partly due to some inherent differences between the biological data used here and images, the latter of which can be represented by a hierarchy of from low- to high-level features requiring a large number of layers or deep neural networks [38]. Given the complexity behind regulatory mechanism of enhancer and promoter, a large number of parameters are still needed for capturing regional/local dependencies and interactions in sequence and epigenomic data (Tables 2 and 3 and Figure 7). Second, among the models for the same data source, probably due to the small number of layers, we noted that adding skip connections did not show a clear advantage over a basic CNN in predicting EPIs (Figure 5a,b). Skip connections as adopted in ResNets were reported to improve prediction performance, partly by better optimizing a deep CNN during the training process [28]. In our work, we observed the optimal number of layers (in a ResNet) at 8 (Figure 1 and Table 3), which was much less than that of a typical ResNet (with 18, 34 or even 1000 layers).

Table 3. Epigenomic model summary.

Block Name	Basic CNN (2 Branches)	Basic CNN (1 Branch)	ResNet with Fully-Connected (fc) Layer	ResNet without Fully-Connected (fc) Layer
Input	$\begin{bmatrix} \text{Enhancer}: 296 \times 22 \\ \text{Promoter}: 196 \times 22 \end{bmatrix}$	$\begin{bmatrix} \text{Enhancer} \\ + \quad : 492 \times 22 \\ \text{Promoter} \end{bmatrix}$	$\begin{bmatrix} \text{Enhancer}: 296 \times 22 \\ \text{Promoter}: 196 \times 22 \end{bmatrix}$	$\begin{bmatrix} \text{Enhancer}: 296 \times 22 \\ \text{Promoter}: 196 \times 22 \end{bmatrix}$
conv1_x	$\begin{bmatrix} 16 \times 1, 256 \\ \text{maxpooling}(2 \times 1) \end{bmatrix} \times 3$	$\begin{bmatrix} 16 \times 1, 256 \\ \text{maxpooling}(2 \times 1) \end{bmatrix} \times 3$	$\begin{bmatrix} 7 \times 1, 64, \text{stride2or1*} \\ \text{maxpooling}(3 \times 1) \end{bmatrix}$	$\begin{bmatrix} 7 \times 1, 64, \text{stride2or1*} \\ \text{maxpooling}(3 \times 1) \end{bmatrix}$
conv2_x			$\begin{bmatrix} 3 \times 1, 256 \\ 3 \times 1, 256 \end{bmatrix} \times 2$	$\begin{bmatrix} 3 \times 1, 128 \\ 3 \times 1, 128 \end{bmatrix} \times 2$
conv3_x			$\begin{bmatrix} 3 \times 1, 256 \\ 3 \times 1, 256 \end{bmatrix} \times 2$	$\begin{bmatrix} 3 \times 1, 128 \\ 3 \times 1, 128 \end{bmatrix} \times 2$
conv4_x				$\begin{bmatrix} 3 \times 1, 128 \\ 3 \times 1, 128 \end{bmatrix} \times 2$
conv5_x				$\begin{bmatrix} 3 \times 1, 128 \\ 3 \times 1, 128 \end{bmatrix} \times 2$
Concatenate branches	$\begin{bmatrix} 512 - \text{dfc} \\ \text{Sigmoid} \end{bmatrix}$	$\begin{bmatrix} \text{Noconcatenation} \\ 512 - \text{dfc} \\ \text{Sigmoid} \end{bmatrix}$	$\begin{bmatrix} \text{avgpooling}(2 \times 1) \\ 800 - \text{dfc} \\ \text{Sigmoid} \end{bmatrix}$	$\begin{bmatrix} \text{avgpooling} \\ \text{Sigmoid} \end{bmatrix}$
# of parameters	8,838,145	12,244,481	5,915,841	1,625,985

* stride 1 for promoter and stride 2 for enhancer.

Finally, besides modeling enhancer or promoter regulatory machineries separately, a large number of parameters was also desirable for characterizing complex interaction patterns between an enhancer and a promoter. We showed that a ResNet without any fully connected layer after the concatenation of the enhancer and promoter branches performed worse than models with fully-connected layers (Table S4), although the result is not significant (Paired t-test p value: 0.2546). In addition, as Figure 7b. demonstrates, 800 fully-connected neurons in the ResNet CNN, corresponding to a higher number of parameters, had the highest validation AUROC. To further illustrate the necessity of having enhancer and promoter as separate branches for CNN models, we also tried inputting aggregated enhancer and promoter epigenomics data at the beginning of the basic CNN model, where the interactions are modeled at the beginning through neural networks (Table 3). As the number of parameters is larger than modeling enhancer and promoter as separate branches, the weighted average AUROC was slightly better but not significant (Table S4; Paired t-test p value: 0.4255), which suggested a more over-parametrized model did not deteriorate the model performance.

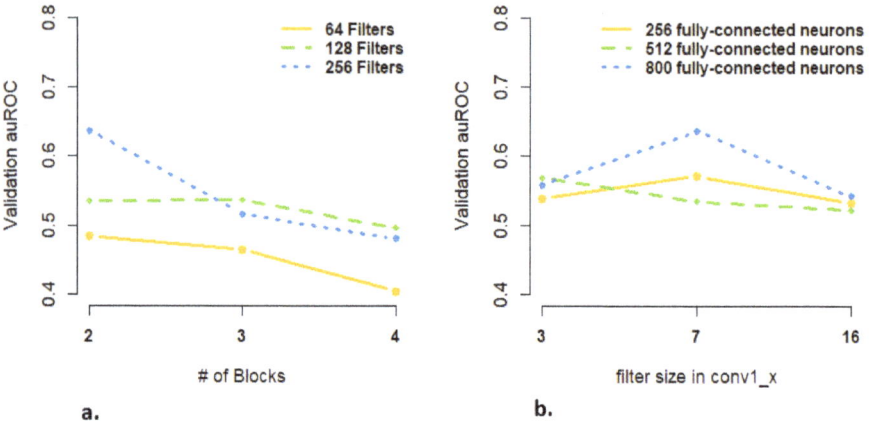

Figure 7. Validation AUROCs for epigenomics ResNet CNNs with various configurations in Table 3: (**a**) the number of ResNet blocks and number of filters for each convolution; (**b**) the filter/kernel size in the first layer and number of fully-connected neurons in the last layer.

Table 4. Epigenomics and sequence model performance.

Epigenomics Model	Mean AUROC	Standard Deviation of AUROC	Sequence Model	Mean AUROC	Standard Deviation of AUROC
Simple convolution/Basic sequence CNN (Zhuang et al., 2019 [14]; original region)	0.500	0.0500	Basic CNN	0.648	0.0704
Attention CNN (central region)	0.529	0.0533	ResNet CNN	0.638	0.0568
ResNet CNN (central region)	0.515	0.0596	Gradient Boosting	0.622	0.0676
Gradient Boosting (central region)	0.494	0.0557	Combined model	0.603	0.0703
Gradient Boosting (original region)	0.499	0.0654			

3.4. Epigenomics Feed-Forward Neural Networks (FNNs) Performed Better than Gradient Boosting

Both FNNs and CNNs had a higher or comparable test AUROC than gradient boosting with either of the data formats (TargetFinder-format and CNN-format) across the 21 test chromosomes for most cell lines (Figure 8 and Table S5). In addition, the training time of FNNs by leveraging GPUs was faster than gradient boosting (e.g., 1–2 min vs. 3–10 min in GB). Little evidences from Table S5 supported that either neural network or gradient boosting was capable of cross-cell-line prediction.

The FNNs performed better than the CNNs in cell lines GM12878 and IMR90 and similar to the CNNs in other two cell lines with slightly lower standard deviations than the CNNs. Although valuable spatial dependency information in a region might be retained in the CNN-format data, the increased data dimension and high noise levels might discount the corresponding benefits. As a side note, the FNNs were still over-parametrized with the input data of dimension only 44 and the training sample size of less than 40,000.

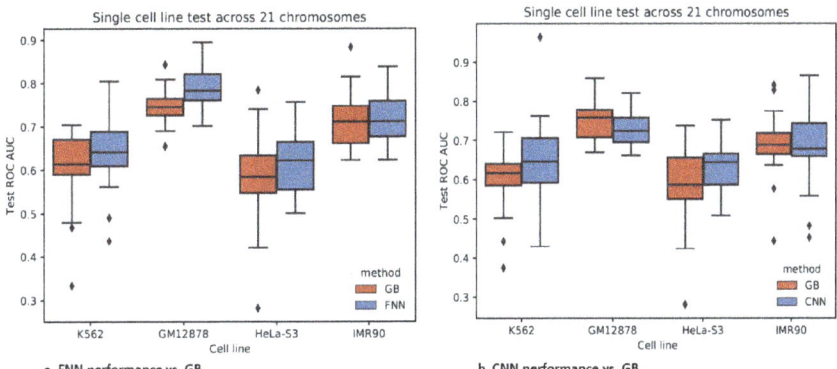

Figure 8. FNN and CNN prediction performance for the same data format as that for GB.

4. Conclusions

Through an extensive evaluation of the use of various neural networks, especially convolutional neural networks (CNNs), on predicting enhancer-promoter interactions (EPIs), we demonstrated that local epigenomic features were more predictive than local sequence data. In contrast to most previous studies on EPI prediction, we reached our conclusions by holding out data from one or more whole chromosomes as training, validation, and test data respectively, avoiding biases associated with random partitioning of enhancer-promoter pairs as training, validation, and test data [19]. We also did not find much predictive gain in integrating local features from the two data sources, perhaps because local sequences were not informative enough for a higher prediction accuracy. We emphasize that, although our findings suggest that local DNA sequence data may not be sufficient to well predict EPIs, a new study has shown some promising results of using mega-base scale sequence data incorporating large-scale genomic context [39]; this is in agreement with improved prediction performance of including not only local epigenomic features of an enhancer and a promoter, but also the window region between them [40]. More studies are warranted.

Supplementary Materials: The following are available online at http://www.mdpi.com/2073-4425/11/1/41/s1, Table S1: Performance of CNNs with varying window and step sizes for the TargetFinder dataset without correct training/validation/test data splitting for cell line GM12878. Table S2: Parameter search grids for CNN models. Table S3: FNN performance comparison between two data formats using chromosome 1 as the test data (with the results for the training data in parentheses). Table S4: Performance summary of additional epigenomics CNN models. Table S5: The single-cell-line and cross-cell-line mean (SD) test AUROCs across each of the 21 test chromosomes for Gradient Boosting (GB) in comparison with the CNNs and FNNs with the same data format.

Author Contributions: Conceptualization, W.P.; methodology, M.X., Z.Z., W.P.; software, M.X., Z.Z.; validation, M.X.; formal analysis, M.X.; investigation, M.X., Z.Z., W.P.; resources, W.P.; data curation, M.X.; writing—original draft preparation, M.X.; writing—review and editing, W.P.; visualization, M.X.; supervision, W.P.; project administration, W.P.; funding acquisition, W.P. All authors have read and agreed to the published version of the manuscript.

Funding: This research was supported by NIH grants R21AG057038, R01HL116720, R01GM113250 and R01HL105397 and R01GM126002, and by the Minnesota Supercomputing Institute.

Conflicts of Interest: The authors declare no conflict of interest. The funders had no role in the design of the study; in the collection, analyses, or interpretation of data; in the writing of the manuscript, or in the decision to publish the results.

References

1. Schoenfelder, S.; Fraser, P. Long-range enhancer-promoter contacts in gene expression control. *Nat. Rev. Genet.* **2019**, *20*, 437–455. [CrossRef] [PubMed]
2. Won, H.; de La Torre-Ubieta, L.; Stein, J.L.; Parikshak, N.N.; Huang, J.; Opland, C.K.; Gandal, M.J.; Sutton, G.J.; Hormozdiari, F.; Lu, D.; et al. Chromosome conformation elucidates regulatory relationships in developing human brain. *Nature* **2016**, *538*, 523. [CrossRef] [PubMed]
3. Wu, C.; Pan, W. Integration of Enhancer-Promoter Interactions with GWAS Summary Results Identifies Novel Schizophrenia-Associated Genes and Pathways. *Genetics* **2018**, *209*, 699–709. [CrossRef] [PubMed]
4. Dekker, J.; Rippe, K.; Dekker, M.; Kleckner, N. Capturing chromosome conformation. *Science* **2002**, *295*, 1306–1311. [CrossRef] [PubMed]
5. Li, G.; Ruan, X.; Auerbach, R.K.; Sandhu, K.S.; Zheng, M.; Wang, P.; Poh, H.M.; Goh, Y.; Lim, J.; Zhang, J.; et al. Extensive promoter-centered chromatin interactions provide a topological basis for transcription regulation. *Cell* **2012**, *148*, 84–98. [CrossRef]
6. Rao, S.S.; Huntley, M.H.; Durand, N.C.; Stamenova, E.K.; Bochkov, I.D.; Robinson, J.T.; Sanborn, A.L.; Machol, I.; Omer, A.D.; Lander, E.S.; et al. A 3D map of the human genome at kilobase resolution reveals principles of chromatin looping. *Cell* **2014**, *159*, 1665–1680. [CrossRef]
7. Sanyal, A.; Lajoie, B.R.; Jain, G.; Dekker, J. The long-range interaction landscape of gene promoters. *Nature* **2012**, *489*, 109. [CrossRef]
8. Cao, Q.; Anyansi, C.; Hu, X.; Xu, L.; Xiong, L.; Tang, W.; Mok, M.T.; Cheng, C.; Fan, X.; Gerstein, M.; et al. Reconstruction of enhancer-target networks in 935 samples of human primary cells, tissues and cell lines. *Nat. Genet.* **2017**, *49*, 1428. [CrossRef]
9. He, B.; Chen, C.; Teng, L.; Tan, K. Global view of enhancer-promoter interactome in human cells. *Proc. Natl. Acad. Sci. USA* **2014**, *111*, E2191–E2199. [CrossRef]
10. Roy, S.; Siahpirani, A.F.; Chasman, D.; Knaack, S.; Ay, F.; Stewart, R.; Wilson, M.; Sridharan, R. A predictive modeling approach for cell line-specific long-range regulatory interactions. *Nucleic Acids Res.* **2015**, *43*, 8694–8712. [CrossRef]
11. Whalen, S.; Truty, R.M.; Pollard, K.S. Enhancer-promoter interactions are encoded by complex genomic signatures on looping chromatin. *Nat. Genet.* **2016**, *48*, 488. [CrossRef] [PubMed]
12. Yang, Y.; Zhang, R.; Singh, S.; Ma, J. Exploiting sequence-based features for predicting enhancer-promoter interactions. *Bioinformatics* **2017**, *33*, I252–I260. [CrossRef] [PubMed]
13. Singh, S.; Yang, Y.; Poczos, B.; Ma, J. Predicting enhancer-promoter interaction from genomic sequence with deep neural networks. *bioRxiv* **2016**, 085241. [CrossRef]
14. Zhuang, Z.; Shen, X.T.; Pan, W. A simple convolutional neural network for prediction of enhancer-promoter interactions with DNA sequence data. *Bioinformatics* **2019**, *35*, 2899–2906. [CrossRef]
15. Geirhos, R.; Rubisch, P.; Michaelis, C.; Bethge, M.; Wichmann, F.A.; Brendel, W. ImageNet-trained CNNs are biased towards texture; increasing shape bias improves accuracy and robustness. *arXiv* **2018**, arXiv:1811.12231.
16. Krizhevsky, A.; Sutskever, I.; Hinton, G.E. Imagenet classification with deep convolutional neural networks. In *Advances in Neural Information Processing Systems*; Pereira, F., Burges, C.J.C., Bottou, L., Weinberger, K.Q., Eds.; Curran Associates, Inc.: Lake Tahoe, NV, USA, 2012; pp. 1097–1105.
17. Luo, X.; Chi, W.; Deng, M. Deepprune: Learning efficient and interpretable convolutional networks through weight pruning for predicting DNA-protein binding. *bioRxiv* **2019**, 729566. [CrossRef]
18. Cao, F.; Fullwood, M.J. Inflated performance measures in enhancer-promoter interaction-prediction methods. *Nat. Genet.* **2019**, *51*, 1196–1198. [CrossRef]
19. Xi, W.; Beer, M.A. Local epigenomic state cannot discriminate interacting and non-interacting enhancer-promoter pairs with high accuracy. *PLoS Comput. Biol.* **2018**, *14*, e1006625. [CrossRef]
20. Li, W.R.; Wong, W.H.; Jiang, R. DeepTACT: Predicting 3D chromatin contacts via bootstrapping deep learning. *Nucleic Acids Res.* **2019**, *47*, e60. [CrossRef]
21. Kong, Y.; Yu, T. A graph-embedded deep feedforward network for disease outcome classification and feature selection using gene expression data. *Bioinformatics* **2018**, *34*, 3727–3737. [CrossRef]
22. Liang, F.; Li, Q.; Zhou, L. Bayesian neural networks for selection of drug sensitive genes. *J. Am. Stat. Assoc.* **2018**, *113*, 955–972. [CrossRef] [PubMed]

23. Bernstein, B.E.; Stamatoyannopoulos, J.A.; Costello, J.F.; Ren, B.; Milosavljevic, A.; Meissner, A.; Kellis, M.; Marra, M.A.; Beaudet, A.L.; Ecker, J.R.; et al. The NIH Roadmap Epigenomics Mapping Consortium. *Nat. Biotechnol.* **2010**, *28*, 1045–1048. [CrossRef] [PubMed]
24. Encode Project Consortium. An integrated encyclopedia of DNA elements in the human genome. *Nature* **2012**, *489*, 57–74. [CrossRef] [PubMed]
25. Quinlan, A.R.; Ira, M.H. BEDTools: A flexible suite of utilities for comparing genomic features. *Bioinformatics* **2010**, *26*, 841–842. [CrossRef] [PubMed]
26. Bahdanau, D.; Cho, K.; Bengio, Y. Neural machine translation by jointly learning to align and translate. *arXiv* **2014**, arXiv:1409.0473.
27. He, K.; Zhang, X.; Ren, S.; Sun, J. Deep residual learning for image recognition. In Proceedings of the IEEE Conference on Computer Vision and Pattern Recognition, Las Vegas, NV, USA, 27–30 June 2016; pp. 770–778.
28. He, K.; Zhang, X.; Ren, S.; Sun, J. Identity mappings in deep residual networks. In Proceedings of the European Conference on Computer Vision, Amsterdam, The Netherlands, 8–16 October 2016; pp. 630–645.
29. Sabour, S.; Frosst, N.; Hinton, G.E. Dynamic routing between capsules. In *Advances in Neural Information Processing Systems*; Guyon, I., Luxburg, U.V., Bengio, S., Wallach, H., Fergus, R., Vishwanathan, S., Garnett, R., Eds.; Curran Associates, Inc.: Long Beach, CA, USA, 2017; pp. 3856–3866.
30. Szegedy, C.; Ioffe, S.; Vanhoucke, V.; Alemi, A.A. Inception-v4, inception-resnet and the impact of residual connections on learning. In Proceedings of the Thirty-First AAAI Conference on Artificial Intelligence, San Francisco, CA, USA, 4–9 February 2017.
31. Ioffe, S.; Szegedy, C. Batch normalization: Accelerating deep network training by reducing internal covariate shift. *arXiv* **2015**, arXiv:1502.03167.
32. Glorot, X.; Bengio, Y. Understanding the difficulty of training deep feedforward neural networks. In Proceedings of the Thirteenth International Conference on Artificial Intelligence and Statistics, Chia Laguna, Italy, 13–15 May 2010; pp. 249–256.
33. He, K.; Zhang, X.; Ren, S.; Sun, J. Delving deep into rectifiers: Surpassing human-level performance on imagenet classification. In Proceedings of the IEEE International Conference on Computer Vision, Santiago, Chile, 7–13 December 2015; pp. 1026–1034.
34. Kingma, D.P.; Ba, J. Adam: A method for stochastic optimization. *arXiv* **2014**, arXiv:1412.6980.
35. Srivastava, N.; Hinton, G.; Krizhevsky, A.; Sutskever, I.; Salakhutdinov, R. Dropout: A Simple Way to Prevent Neural Networks from Overfitting. *J. Mach. Learn. Res.* **2014**, *15*, 1929–1958.
36. Jing, F.; Zhang, S.; Cao, Z.; Zhang, S. An integrative framework for combining sequence and epigenomic data to predict transcription factor binding sites using deep learning. *IEEE ACM Trans. Comput. Biol. Bioinform.* **2019**. [CrossRef]
37. Nair, S.; Kim, D.S.; Perricone, J.; Kundaje, A. Integrating regulatory DNA sequence and gene expression to predict genome-wide chromatin accessibility across cellular contexts. *Bioinformatics* **2019**, *35*, i108–i116. [CrossRef]
38. LeCun, Y.; Bengio, Y.; Hinton, G. Deep learning. *Nature* **2015**, *521*, 436–444. [CrossRef] [PubMed]
39. Schwessinger, R.; Gosden, M.; Downes, D.; Brown, R.; Telenius, J.; Teh, Y.W.; Lunter, G.; Hughes, J.R. DeepC: Predicting chromatin interactions using megabase scaled deep neural networks and transfer learning. *bioRxiv* **2019**, 724005. [CrossRef]
40. Whalen, S.; Pollard, K.S. Reply to 'Inflated performance measures in enhancer-promoter interaction-prediction methods'. *Nat. Genet.* **2019**, *51*, 1198–1200. [CrossRef] [PubMed]

© 2019 by the authors. Licensee MDPI, Basel, Switzerland. This article is an open access article distributed under the terms and conditions of the Creative Commons Attribution (CC BY) license (http://creativecommons.org/licenses/by/4.0/).

Article

Penalized Variable Selection for Lipid–Environment Interactions in a Longitudinal Lipidomics Study

Fei Zhou [1], Jie Ren [1], Gengxin Li [2], Yu Jiang [3], Xiaoxi Li [1], Weiqun Wang [4] and Cen Wu [1,*]

[1] Department of Statistics, Kansas State University, Manhattan, KS 66506, USA; feiz@ksu.edu (F.Z.); jieren@ksu.edu (J.R.); xiaoxili@ksu.edu (X.L.)
[2] Department of Mathematics and Statistics, University of Michigan Dearborn, Dearborn, MI 48128, USA; gengxinl@umich.edu
[3] Division of Epidemiology, Biostatistics, and Environmental Health, School of Public Health, University of Memphis, Memphis, TN 38111, USA; yjiang4@memphis.edu
[4] Department of Food, Nutrition, Dietetics and Health, Kansas State University, Manhattan, KS 66506, USA; wwang@ksu.edu
* Correspondence: wucen@ksu.edu; Tel.: +1-785-532-2231

Received: 7 November 2019; Accepted: 26 November 2019; Published: 3 December 2019

Abstract: Lipid species are critical components of eukaryotic membranes. They play key roles in many biological processes such as signal transduction, cell homeostasis, and energy storage. Investigations of lipid–environment interactions, in addition to the lipid and environment main effects, have important implications in understanding the lipid metabolism and related changes in phenotype. In this study, we developed a novel penalized variable selection method to identify important lipid–environment interactions in a longitudinal lipidomics study. An efficient Newton–Raphson based algorithm was proposed within the generalized estimating equation (GEE) framework. We conducted extensive simulation studies to demonstrate the superior performance of our method over alternatives, in terms of both identification accuracy and prediction performance. As weight control via dietary calorie restriction and exercise has been demonstrated to prevent cancer in a variety of studies, analysis of the high-dimensional lipid datasets collected using 60 mice from the skin cancer prevention study identified meaningful markers that provide fresh insight into the underlying mechanism of cancer preventive effects.

Keywords: GEE; lipid–environment interaction; longitudinal lipidomics study; penalized variable selection

1. Introduction

Longitudinal data are frequently observed in a diversity of scientific research areas, including economics, biomedical studies, and clinical trials. A common characteristic of the longitudinal data is that the same subject is measured repeatedly over a certain period of time; thus, the repeated measurements are correlated. Many modeling techniques have been proposed to accommodate the multivariate correlated nature of the data [1,2]. The emergence of new types of data has brought constant challenges to the development of novel statistical methods for longitudinal studies. One representative example is the high-dimensional data where the number of variables is much larger than the sample size. As penalization has been demonstrated as an effective way for conducting variable selection in linear and generalized linear models with a univariate response [3,4], substantial efforts have been devoted to developing penalized variable selection methods with longitudinal responses, such as [5–7], among many others.

This study was partially motivated by overcoming the limitations of existing penalization methods in order to analyze the high-dimensional lipidomics data from longitudinal studies. Lipids are a broad group of biomolecules in eukaryotic membranes, involved in various critical biological roles such as energy storage, cellular membrane structure, or cell signaling and homeostasis [8–11]. Lipid metabolism has been found to be associated with several diseases, especially chronic diseases such as diabetes, cancer, inflammatory disease, and Alzheimer [12–14].

The lipid data were obtained from our previous work on the lipid changes in weight controlled CD-1 mice [15]. In the current study, the phenotype of interest is the body weight of experimental animals, which was measured every week for 10 weeks. The environmental factor was exercise and/or dietary restriction, which had four different levels, control (ad libitum feeding and sedentary), AE (exercise and ad libitum feeding), PE (exercise and pair feeding), and DCR (sedentary and 20% dietary calorie restriction). Both triacylglycerol (TG) and diacylglycerol (DG) profiles in the plasma were measured using electrospray ionization MS/MS [15]. Here, we focused on the DG profiles and treated them as lipid factors. Besides the lipid main effects, we were particularly interested in investigating the interactions between lipids and environment/treatment effects, which will shed novel insight in the understanding of weight changes in a longitudinal setting beyond studies solely focusing on the main lipidomics effects. With the control as the baseline, we created a group of three dummy variables to represent the four levels of the treatment factor that can be treated as environmental factors in general. The product between the dummy variable group and lipid denotes the lipid–environment interactions. The formulation of the interaction group in our study shared the spirit of group LASSO, which was primarily motivated by the selection of important dummy variable groups from ANOVA problems [16]. As the total number of main and interaction effects was much larger than the sample size, penalized variable selection was a natural choice to identify the important subset of effects. Such methods for G×E interactions, including [17,18], however, cannot be adopted for the longitudinal studies.

On the other hand, existing penalization methods in longitudinal studies have been mostly developed for the identification of important main effects only. For instance, Wang et al. [5] proposed the penalized generalized estimating equation (PGEE) to select predictors that are associated with the longitudinal response. Ma et al. [6] considered the selection of important predictors and estimation of non-parametric effects through splines for repeated measures data. Cho and Qu [7] developed a penalized quadratic inference function (PQIF) method to conduct variable selection on main effects. Fan et al. [19] developed robust variable selection through a penalized robust estimating equation to incorporate the correlation structure for repeated measurements. These studies have ignored the interaction effects and cannot be adopted to analyze our data directly. In addition, our limited search also suggests that user-friendly software packages for variable selection methods in longitudinal studies have been relatively underdeveloped. For penalization methods, only two R packages (PGEE and pgee.mixed) are available, and both packages have focused on the selection of important main effects. The codes for most studies in this area have not even been made publicly available.

To accommodate simultaneously the selection of individual and group structure corresponding to the main lipid effect and interaction effect respectively, we propose a novel penalized variable selection method for longitudinal clustered data. Our method significantly advances the existing penalization methods by considering the interaction effects. Through incorporating the group structure, selection of both main and interaction effects can be efficiently conducted within the generalized estimating equation framework [20]. Furthermore, to facilitate fast computation and reproducible research, we implement the proposed and benchmark methods in the R package (interep https://cran.r-project.org/package=interep) [21]. The software package is open-source, and the core module has been developed in C++. The advantage of our method over alternatives has been demonstrated in extensive simulation studies. Analysis of the motivating dataset yields findings with important implications.

2. Materials and Methods

2.1. Data and Model Settings

Consider a longitudinal study with n subjects and k_i observations measured repeatedly across time for the ith subject ($1 \leqslant i \leqslant n$). Without loss of generality, we set $k_i = k$. Y_{ij} denotes the response for the ith subject at time j ($1 \leqslant j \leqslant k$). $X_{ij} = (X_{ij1}, ..., X_{ijp})^\top$ is the p-dimensional vector of lipid factors. In our study, $E_{ij} = (E_{ij1}, ..., E_{ijq})^\top$ denotes the q-dimensional treatment factor. Suppose that the lipid factors, treatment factors, and their interactions are associated with the longitudinal phenotype through the following model:

$$Y_{ij} = \beta_0 + E_{ij}^\top \beta_1 + X_{ij}^\top \beta_2 + (X_{ij} \otimes E_{ij})^\top \beta_3 + \epsilon_{ij} = Z_{ij}^\top \beta + \epsilon_{ij} \quad (1)$$

where $\beta = (\beta_0, \beta_1^\top, \beta_2^\top, \beta_3^\top)^\top$ and $Z_{ij} = c(1, E_{ij}^\top, X_{ij}^\top, (X_{ij} \otimes E_{ij})^\top)^\top$ are $(1 + q + p + pq)$-dimensional vectors that represent all the main and interaction effects. The interactions between lipids and treatment factors are modeled through $X_{ij} \otimes E_{ij}$, the Kronecker product of the p-dimensional vector X_{ij}, and the q-dimensional vector E_{ij} within the following form:

$$X_{ij} \otimes E_{ij} = [X_{ij1}E_{ij1}, X_{ij1}E_{ij2}, ..., X_{ij1}E_{ijq}, X_{ij2}E_{ij1}, ..., X_{ijp}E_{ijq}]^\top$$

which is a pq-dimensional vector. β_0 is the intercept. β_1, β_2, and β_3 are unknown coefficient vectors of dimensions q, p, and pq, respectively. We assume that the observations are dependent within the same subject, and independent if they are from different subjects. $\epsilon_i = (\epsilon_{i1}, ..., \epsilon_{ik_i})^\top$ follows a multivariate normal distribution $N_k(0, \Sigma_i)$, with Σ_i as the covariance matrix for the repeated measure of the ith subject across the k time points.

2.2. Generalized Estimating Equations

The joint likelihood function for longitudinally clustered response Y_{ij} is generally difficult to specify. Liang and Zeger [20] developed the generalized estimating equations (GEE) method to account for the intra-cluster correlation. It models the marginal instead of the conditional distribution given the previous observations and only requires a working correlation structure for Y_{ij} to be specified.

The first two marginal moments of Y_{ij} are denoted by $E(Y_{ij}) = \mu_{ij} = Z_{ij}^T \beta$ and $Var(Y_{ij}) = v(\mu_{ij})$, respectively, where v is a known variance function. Then, the estimating equation for β is defined as:

$$\sum_{i=1}^{n} \frac{\partial \mu_i(\beta)}{\partial \beta} V_i^{-1}(Y_i - \mu_i(\beta)) = 0, \quad (2)$$

where $\mu_i(\beta) = (\mu_{i1}(\beta), ..., \mu_{ik}(\beta))^\top$, $Y_i = (Y_{i1}, ..., Y_{ik})^\top$ and V_i is the covariance matrix of Y_i. The first term in (2), $\frac{\partial \mu_i(\beta)}{\partial \beta}$, reduces to $Z_i = (Z_{i1}, ..., Z_{ik})^\top$, which corresponds to the $k \times (1 + q + p + pq)$ matrix of the main and interaction effects.

V_i is often unknown in practice and difficult to estimate especially when the number of variance components is large. In GEE, the covariance matrix V_i is specified through a simplified working correlation matrix $R(\eta)$ as $V_i = A_i^{\frac{1}{2}} R(\eta) A_i^{\frac{1}{2}}$, with the diagonal marginal variance matrix $A_i = \text{diag}\{Var(Y_{i1}), ..., Var(Y_{ik})\}$. $R(\eta)$ is characterized by a finite-dimensional nuisance parameter η. Commonly adopted correlation structures for $R(\eta)$ can be independent, AR(1), and exchangeable, among others. Liang and Zeger [20] showed that if η can be consistently estimated, the GEE estimator of the regression coefficient is consistent. Furthermore, the consistency holds even when the working correlation structure is misspecified.

2.3. Penalized Identification

When the dimensionality of lipid factors is high, the total number of main and interaction effects is even higher. However, only a small subset of important effects is associated with the phenotype, which naturally leads to a variable selection problem. Penalized GEE based methods, including Wang et al. [5] and Ma et al. [6], have been proposed for conducting selection under correlated longitudinal responses. However, those published studies focus on the main effects and ignore the interactions. As shown in (1), the lipid–environment interactions are modeled on the group level, that is the interaction between all the q treatment factors and the hth lipidomics measurement ($1 \leqslant h \leqslant p$). Such a group structure cannot be accommodated by variable selection methods from existing longitudinal studies. This fact motivates us to develop a method for the interaction analysis of repeated measures data, termed as interep, with the following penalized generalized estimating equation:

$$Q(\beta) = U(\beta) - \sum_{g=1}^{p} \rho'(|\beta_{2g}|; \lambda_1, \gamma) \text{sign}(\beta_{2g}) - \sum_{h=1}^{p} \rho'(||\beta_{3h}||_{\Sigma_h}; \sqrt{q}\lambda_2, \gamma), \quad (3)$$

where $U(\beta)$ is the score equation in GEE and $\rho'(\cdot)$ is the first derivative of the minimax concave penalty (MCP) [22]. Since the environmental factors are usually of low dimension and are predetermined as important ones, they are not subject to penalized selection. $U(\beta)$ is defined as:

$$U(\beta) = \sum_{i=1}^{n} Z_i^T V_i^{-1}(Y_i - \mu_i(\beta)),$$

and the MCP can be expressed as:

$$\rho(t; \lambda, \gamma) = \lambda \int_0^t (1 - \frac{x}{\gamma \lambda})_+ dx,$$

where λ is the tuning parameter and γ is the regularization parameter. The first derivative function of the MCP penalty is:

$$\rho'(t; \lambda, \gamma) = (\lambda - \frac{t}{\gamma}) I(t \leq \gamma \lambda).$$

MCP can be adopted for the regularized selection on both individual and group level effects. It is fast, continuous, and nearly unbiased [22].

In (3), the vector $\beta_2 = (\beta_{21}, ..., \beta_{2p})^\top$ denotes the regression parameters for all the p lipid factors. $\beta_3 = (\beta_{31}^\top, ..., \beta_{3p}^\top)^\top$, which denotes the regression parameters for lipid–environment interactions, is a vector of length pq. β_{3h} is a vector of length q ($h = 1, 2, ..., p$), corresponding to the interactions between the hth lipid feature and the environment factors. With the control as the baseline, the environment factors have been formulated as a group of dummy variables. With high-dimensional main and interaction effects, penalization is critical for the identification of important effects out of the large number of candidates. In the penalized generalized estimating equation (3), the first penalty term adopts MCP directly to conduct the selection of main lipid effects on the individual level. The second penalty, in the forms of group MCP, imposes shrinkage on the product between the lipid factors and dummy variable group, which corresponds to the lipid–environment interactions. The group level selection of interaction effects is consistent with the mechanism of creating the dummy variable group of environmental factors. Note that such a rationale of formulating the penalized generalized estimating Equation (3) is deeply rooted in group LASSO [16].

In particular, λ_1 and λ_2 in (3) are tuning parameters. $\rho'(||\beta_{3h}||_{\Sigma_h}; \sqrt{q}\lambda_2, \gamma)$ is the group MCP penalty that corresponds to the interactions between the hth ($h = 1, 2, ..., p$) lipid factor and the q environment factors. The empirical norm $||\beta_{3h}||_{\Sigma_h}$ is defined as: $||\beta_{3h}||_{\Sigma_h} = (\beta'_{3h} \Sigma_h \beta_{3h})^{1/2}$ with $\Sigma_h = n^{-1} B_h^\top B_h$. $B_h = Z[, (2 + q + p + (h-1) \times q) : (1 + q + p + h \times q)]$, and it contains the q columns in Z that correspond to the interactions from the hth lipid factor with the q environment factors.

A variety of penalized variable selection methods for high-dimensional longitudinal data have been developed in the past two decades for analyzing high-dimensional omics data, such as gene expressions, single nucleotide polymorphisms (SNPs), and copy number variations (CNVs) [5,6]. However, lipidomics data have been rarely investigated by using high-dimensional variable selection methods. We developed a package, (interep https://cran.r-project.org/package=interep)

that incorporates our recently developed penalization procedures to conduct interaction analysis for high-dimensional lipidomics data with repeated measurements [21].

Remark: The uniqueness of the proposed study lies in accounting for the group structure of lipid–environment interactions through penalized identification. Therefore, the main lipid effects and lipid–environment interactions are penalized on individual and group levels, separately, which leads to a formulation of both MCP and group MCP penalties. Although our model has been motivated from a specific lipidomics profiling study in weight controlled mice [15], it can be readily extended to accommodate more general cases in interaction studies where the environmental factors are not dummy variables formulated from the ANOVA setting. In such a case, for each lipid factor, the main lipid effects and lipid–environment interactions form a group, with the leading component of the group being a vector of 1s. As not all the effects in the group are expected to be associated with the phenotype, a sparse group type of variable selection is demanded. Such a formulation has been investigated in survival analysis [23], but not in longitudinal studies yet. With a simple modification of our model to penalize the main and interaction effects on the individual and group level simultaneously, the proposed one becomes a penalized sparse group GEE model and can be adopted to handle general environmental factors in high-dimensional cancer genomics studies.

2.4. Computational Algorithms

We developed an efficient Newton–Raphson type of algorithm to obtain the penalized estimate $\hat{\beta}$. Starting with an initialized value, we can solve the penalized GEE iteratively. The estimated $\hat{\beta}^{(d+1)}$ in the $(d+1)$th iteration can be solved as:

$$\hat{\beta}^{(d+1)} = \hat{\beta}^{(d)} + [T^{(d)} + nW^{(d)}]^{-1}[U^{(d)} - nW^{(d)}\hat{\beta}^{(d)}], \qquad (4)$$

where $U^{(d)}$ is the score function expressed in terms of $\hat{\beta}^{(d)}$ at the dth iteration and $T^{(d)}$ is the corresponding first derivative function of $U^{(d)}$:

$$T^{(d)} = \sum_{i=1}^{n} Z_i^T V_i^{-1} Z_i,$$

which is also a function of $\hat{\beta}^{(d)}$. The MCP penalty was imposed on both the individual level (main lipid effects) and group level (lipid–environment interactions). Therefore, $W^{(d)}$ is a diagonal matrix that contains the first derivative of the MCP penalty for the lipid factors and the first derivative of the group MCP penalty for the lipid–environment interactions. We define $W^{(d)}$ as:

$$W^{(d)} = \mathrm{diag}\{\underbrace{0, \ldots, 0}_{1+q}, \frac{\rho'(|\hat{\beta}_{21}^{(d)}|; \lambda_1, \gamma)}{\epsilon + |\hat{\beta}_{21}^{(d)}|}, \ldots, \frac{\rho'(|\hat{\beta}_{2p}^{(d)}|; \lambda_1, \gamma)}{\epsilon + |\hat{\beta}_{2p}^{(d)}|}, \frac{\rho'(||\hat{\beta}_{31}^{(d)}||_{\Sigma_1}; \sqrt{q}\lambda_2, \gamma)}{\epsilon + ||\hat{\beta}_{31}^{(d)}||_{\Sigma_1}}, \ldots, \frac{\rho'(||\hat{\beta}_{3p}^{(d)}||_{\Sigma_p}; \sqrt{q}\lambda_2, \gamma)}{\epsilon + ||\hat{\beta}_{3p}^{(d)}||_{\Sigma_p}}\},$$

where ϵ is a small positive number set to 10^{-6} to avoid the numerical instability when the denominator is zero. The first $(1 + q)$ elements on the diagonal of W are zero, suggesting that there is no shrinkage imposed on the coefficients for the intercept and the environmental factors. We can use $nW\hat{\beta}$ and nW to approximate the first derivative function of MCP in the penalized score equation and the second derivative function of the MCP penalty, respectively. Given a fixed tuning parameter, the regression parameter $\hat{\beta}^{(d+1)}$ can be updated iteratively till convergence. The stopping criterion is that the L1 norm for the L1 difference between two consecutive iterations is less than 10^{-3}, and convergence can usually be achieved within 10 iterations.

There are two tuning parameters λ_1 and λ_2 and a regularization parameter γ. λ_1 controls the sparsity of lipid factors, and λ_2 determines sparsity among lipid–environment interactions. We chose the optimal tuning parameters λ_1 and λ_2 using five-fold cross-validation in both the simulation study and real data analysis. The regularization parameter γ was obtained via a data driven approach. In our numerical study, we examined a sequence of values, such as 1.8, 3, 4.5, 6, and 10, suggested by published studies, and found that the results were not sensitive to the choice of the value of γ, and then set the value at 3. We split the dataset into five equally sized subsets and took four of them as the training dataset, leaving the last subset as the testing dataset. The penalized estimates were obtained from the training data, and then, prediction performance was evaluated on the testing data. A joint search over a two-dimensional grid of (λ_1, λ_2) was conducted to find the optimal pair of tuning parameters.

Given fixed tuning parameters, we implemented the algorithm as follows:

(1) Set the initial coefficient vector $\beta^{(0)}$ using LASSO;
(2) Update $\beta^{(d+1)}$ using Equation (4) at the $(d+1)$th iteration;
(3) Repeat Step (2) until the convergence criterion is satisfied.

In our study, we considered the methods considering both lipid main effects and lipid–environment interactions with exchangeable working correlation (A1), AR(1) working correlation (A2), and independence working correlation (A3). For comparison with the methods that cannot accommodate the identification of lipid–environment interactions, we also included A4–A6, which incorporate the exchangeable, AR(1), and independence working correlation, respectively. The alternative methods A4–A6 do not ignore the interaction effects. Instead, they treat the interaction effects individually, so the group structure considered in A1–A3 does not exist. We computed the CPU running time for 100 replicates of simulated lipidomics data with $n = 250, \rho = 0.8, p = 75$ (with a total dimension of 304) and fixed tuning parameters on a regular laptop for A1–A6, which can be implemented using our developed package: (interep https://cran.r-project.org/package=interep) [21]. The CPU running time in seconds was 48.8 (A1), 40.2 (A2), 29.0 (A3), 49.3 (A4), 39.7 (A5), and 27.9 (A6), respectively.

3. Results

3.1. Simulation

We evaluated the performance of all six methods (A1–A6) through extensive simulation studies. Among them, A1–A3 were developed for accommodating the interaction structures with different working correlations, while A4–A6 were only focused on the identification of main effects so the structure of the group level interaction effects were not respected. Note that there are existing studies that can also achieve the selection of main effects in longitudinal studies. For example, Wang et al. [5] adopted the smoothly clipped absolute deviation (SCAD) penalty for conducting the selection of main effects. Since the MCP is incorporated as the baseline penalty in A1–A3, A4–A6 have thus been developed based on MCP and used as benchmark methods for comparison.

The responses were generated from the model (2) with sample size $n = 250$ and 500. The number of time points k was set to five. The dimensions for lipid factors X_{ij} were $p = 75$, 150 and 300. With $q = 3$ for E_{ij}, we first simulated a vector of length n from the standard normal distribution. A group of three binary dummy variables for environmental factors could then be generated after dichotomizing the vector at the 30th and 70th percentiles. In addition, the lipids were simulated from a multivariate normal distribution with mean zero and the AR1 covariance matrix with marginal variance one and auto-correlation coefficient 0.5. We simulated the random error ϵ from a multivariate normal distribution by assuming a zero mean vector and an AR1 covariance structure with $\rho = 0.5$ and 0.8. Note that when considering the interactions, the actual dimensionality was much larger than p. For instance, given $n = 250$, $p = 150$, and $q = 3$, the total dimension for all the main and interaction effects was 604.

The coefficients were simulated from $U[0.4, 0.8]$ for 17 nonzero effects, consisting of the intercept, 3 environmental dummy variables, 4 lipid main effects, and 3 groups of lipid–environment interactions

(9 interaction effects). We generated 100 replicates for the four settings: (1) $n = 250$ and $p = 75$, (2) $n = 250$ and $p = 150$, (3) $n = 500$ and $p = 150$, and (4) $n = 500$ and $p = 300$. All the rest of the coefficients were set to zero. For each setting, we considered two correlation coefficients ($\rho = 0.5$ and 0.8) for the random error. The number of true positives (TP) and false positives (FP) was recorded.

In addition to identification results, we also calculated the estimation accuracy in terms of the difference between estimated and true coefficients. In particular, the mean squared error corresponding to the true nonzero coefficients and true zero coefficients (for noisy effects) were termed as MSE and NMSE, respectively. The total mean squared error for the coefficient vector, or TMSE, is computed as:

$$\text{TMSE} = \frac{1}{100} \sum_{r=1}^{100} ||\hat{\beta}^{(r)} - \beta||^2 / p_\beta$$

where p_β is the dimension of β and $\hat{\beta}^{(r)}$ is the estimated value of β in the rth simulated dataset. MSE and NMSE were calculated in a similar way as for TMSE.

Identification results of the six methods (A1–A6) are tabulated in Tables 1–4. In general, A1–A3, which account for both the lipid main effects and lipid–environment interactions, had better performance than A4–A6, which only accommodated the main effects. For example, in Table 1, given $n = 250$, $\rho = 0.5$, $p = 75$, the actual dimension is 304. A1 identified 14.5 (sd 1.9) nonzero effects out of all the 17 true positives, with a relatively small number of false positives of 4.8 (sd 3.1). On the other hand, A4 identified a smaller number of true positives, 1.3 (sd 1.5), with a larger number of false positives, 6.6 (sd 4.2). Among the identified effects, A1 identified 7.4 (sd 1.5) interactions, with 3.1 (sd 2.6) false positives. A4 identified a smaller TP of 6.1 (sd 1.1) and a higher FP of 5.1 (sd 3.3) of the lipid–environment interactions. We could observe that the difference in identification performance between A1 and A4 came mainly from the interaction effects, which was due to the fact that A4 could not accommodate the group level selection corresponding to the lipid–environment interactions. As the dimension increased, A1 outperformed A4 more significantly. For instance, in Table 4, the overall dimension for $n = 500$, $\rho = 0.8$, $p = 300$ is 1204. A1 had a TP of 15.9 (sd 1.2) and an FP of 3 (sd 2.6), while A4 had a smaller TP 14.5 (sd 1.2) and a higher FP 4.5 (sd 3.0). Figures 1 and 2 are plotted based on the identification results from Tables 1–4. We can observe that overall, A1–A3 outperformed A4–A6 with a higher TP and a lower FP under each setting.

Table 1. Identification results for $n = 250$, $p = 75$ with an actual dimension of 304.

$n = 250$	$p = 75$	Overall		Main		Interaction	
		TP	FP	TP	FP	TP	FP
$\rho = 0.5$	A1	14.5(1.9)	4.8(3.1)	7.2(0.8)	1.7(1.2)	7.4(1.5)	3.1(2.6)
	A2	14.7(1.8)	5.0(3.2)	7.2(0.9)	1.7(1.3)	7.5(1.4)	3.2(2.6)
	A3	14.7(1.7)	5.0(3.3)	7.2(0.8)	1.8(1.4)	7.6(1.3)	3.2(2.6)
	A4	13.3(1.5)	6.6(4.2)	7.2(0.7)	1.6(1.4)	6.1(1.1)	5.1(3.3)
	A5	13.3(1.5)	6.8(4.4)	7.2(0.8)	1.7(1.4)	6.1(1.1)	5.2(3.5)
	A6	13.3(1.5)	7.3(4.7)	7.2(0.8)	1.8(1.5)	6.1(1.1)	5.5(3.7)
$\rho = 0.8$	A1	13.7(2.3)	4.1(2.8)	7.2(0.8)	1.5(1.0)	6.5(2.1)	2.7(2.4)
	A2	13.9(2.4)	4.1(2.8)	7.2(0.8)	1.5(1.0)	6.6(2.1)	2.7(2.4)
	A3	14.2(2.3)	4.5(2.9)	7.2(0.7)	1.6(1.0)	7.0(2.2)	2.9(2.5)
	A4	12.9(1.9)	5.5(2.7)	7.2(0.7)	1.1(1.0)	5.6(1.6)	4.5(2.3)
	A5	12.9(1.9)	5.8(2.9)	7.2(0.7)	1.1(0.9)	5.7(1.6)	4.7(2.5)
	A6	13.0(1.8)	6.5(3.5)	7.2(0.7)	1.2(0.9)	5.8(1.4)	5.5(3.2)

Mean (sd) based on 100 replicates. A1–A3: methods accommodating the lipid–environment interactions with exchangeable, AR(1), and independence working correlations, respectively. A4–A6: methods not accommodating the lipid–environment interactions with exchangeable, AR(1), and independence working correlations, respectively.

Table 2. Identification results for $n = 250$, $p = 150$ with an actual dimension of 604.

$n = 250$	$p = 150$	Overall		Main		Interaction	
		TP	FP	TP	FP	TP	FP
	A1	13.9(2.3)	5.0(3.0)	7.2(0.7)	1.7(1.1)	6.7(2.0)	3.3(2.6)
	A2	14.0(2.2)	5.0(3.0)	7.2(0.7)	1.7(1.1)	6.8(1.9)	3.3(2.6)
$\rho = 0.5$	A3	14.4(2.2)	5.1(3.2)	7.3(0.7)	1.8(1.2)	7.1(1.9)	3.3(2.8)
	A4	12.9(1.9)	5.7(2.5)	7.3(0.8)	1.4(0.9)	5.6(1.5)	4.4(2.3)
	A5	13.0(1.8)	5.9(2.6)	7.2(0.8)	1.4(0.9)	5.7(1.4)	4.5(2.3)
	A6	13.0(1.8)	6.4(2.7)	7.2(0.8)	1.4(1.0)	5.8(1.5)	5.0(2.5)
	A1	13.5(2.0)	5.3(3.0)	7.2(0.9)	2.1(1.2)	6.3(1.9)	3.2(2.4)
	A2	13.5(2.0)	5.4(3.2)	7.2(0.9)	2.2(1.3)	6.3(1.9)	3.2(2.5)
$\rho = 0.8$	A3	13.4(2.1)	6.0(3.0)	7.1(0.9)	2.4(1.3)	6.2(1.9)	3.6(2.7)
	A4	12.5(1.9)	7.6(3.3)	7.3(0.7)	1.8(1.2)	5.2(1.7)	5.7(2.7)
	A5	12.6(1.8)	7.8(3.4)	7.3(0.7)	1.9(1.2)	5.3(1.6)	5.9(2.8)
	A6	12.6(1.8)	8.4(4.1)	7.3(0.8)	1.9(1.2)	5.4(1.7)	6.5(3.6)

Mean (sd) based on 100 replicates. A1–A3: methods accommodating the lipid–environment interactions with exchangeable, AR(1), and independence working correlations, respectively. A4–A6: methods not accommodating the lipid–environment interactions with exchangeable, AR(1), and independence working correlations, respectively.

Table 3. Identification results for $n = 500$, $p = 150$ with an actual dimension of 604.

$n = 500$	$p = 150$	Overall		Main		Interaction	
		TP	FP	TP	FP	TP	FP
	A1	15.7(1.4)	2.7(1.9)	7.7(0.5)	1.3(0.7)	8.0(1.4)	1.4(1.7)
	A2	15.8(1.3)	2.7(2)	7.7(0.5)	1.3(0.7)	8.1(1.3)	1.3(1.8)
$\rho = 0.5$	A3	16.2(1.2)	2.7(1.9)	7.8(0.4)	1.3(0.8)	8.4(1.2)	1.3(1.6)
	A4	14.7(1.0)	2.5(1.7)	7.8(0.4)	0.9(0.8)	6.9(1.0)	1.6(1.4)
	A5	14.7(1.1)	2.6(1.7)	7.8(0.4)	0.9(0.7)	6.9(1.0)	1.7(1.4)
	A6	14.9(1.0)	2.7(2.0)	7.8(0.4)	0.8(0.7)	7.0(0.9)	1.8(1.6)
	A1	15.5(1.7)	3.0(2.9)	7.7(0.6)	1.1(0.8)	7.9(1.5)	1.9(2.2)
	A2	15.4(1.7)	2.9(2.8)	7.7(0.6)	1.1(0.8)	7.8(1.5)	1.8(2.2)
$\rho = 0.8$	A3	15.7(1.6)	2.6(2.6)	7.7(0.5)	1.2(0.9)	8.0(1.4)	1.4(2.1)
	A4	14.8(1.4)	3.7(1.8)	7.5(0.6)	1.2(0.7)	7.2(1.2)	2.5(1.5)
	A5	14.7(1.3)	3.6(1.9)	7.5(0.5)	1.1(0.7)	7.2(1.2)	2.5(1.5)
	A6	15.0(1.3)	3.8(1.9)	7.7(0.6)	1.1(0.7)	7.4(1.1)	2.7(1.6)

Mean (sd) based on 100 replicates. A1–A3: methods accommodating the lipid–environment interactions with exchangeable, AR(1), and independence working correlations, respectively. A4–A6: methods not accommodating the lipid–environment interactions with exchangeable, AR(1), and independence working correlations, respectively.

Table 4. Identification results for $n = 500$, $p = 300$ with an actual dimension of 1204.

$n = 500$	$p = 300$	Overall		Main		Interaction	
		TP	FP	TP	FP	TP	FP
$\rho = 0.5$	A1	16.1(1.2)	3.2(2.4)	7.6(0.6)	1.4(0.8)	8.5(1.0)	1.8(2.2)
	A2	16.3(1.1)	3.2(2.4)	7.7(0.5)	1.4(0.8)	8.5(0.9)	1.8(2.2)
	A3	16.3(1)	2.9(2.2)	7.8(0.5)	1.4(0.8)	8.6(0.8)	1.5(1.9)
	A4	14.8(0.8)	2.9(2.1)	7.8(0.4)	1.0(0.8)	7.0(0.8)	1.9(1.7)
	A5	14.8(0.9)	3.1(2.3)	7.8(0.4)	1.0(0.8)	7.0(0.8)	2.0(1.9)
	A6	14.9(0.9)	3.3(2.6)	7.8(0.4)	1.0(0.8)	7.1(0.9)	2.3(2.1)
$\rho = 0.8$	A1	15.9(1.2)	3(2.6)	7.6(0.5)	1.5(0.8)	8.3(1.1)	1.5(2.2)
	A2	15.9(1.3)	3.0(2.7)	7.6(0.5)	1.5(0.9)	8.2(1.1)	1.5(2.2)
	A3	15.8(1.4)	3.1(2.8)	7.7(0.5)	1.6(1.0)	8.1(1.2)	1.6(2.2)
	A4	14.5(1.2)	4.5(3.0)	7.8(0.6)	1.0(0.7)	6.8(1.0)	3.5(2.6)
	A5	14.5(1.2)	4.7(3.3)	7.8(0.6)	1.1(0.8)	6.7(0.9)	3.6(2.9)
	A6	14.5(1.1)	4.9(3.6)	7.8(0.6)	1.0(0.8)	6.7(0.8)	3.8(3.3)

Mean (sd) based on 100 replicates. A1–A3: methods accommodating the lipid–environment interactions with exchangeable, AR(1), and independence working correlations, respectively. A4–A6: methods not accommodating the lipid–environment interactions with exchangeable, AR(1), and independence working correlations, respectively.

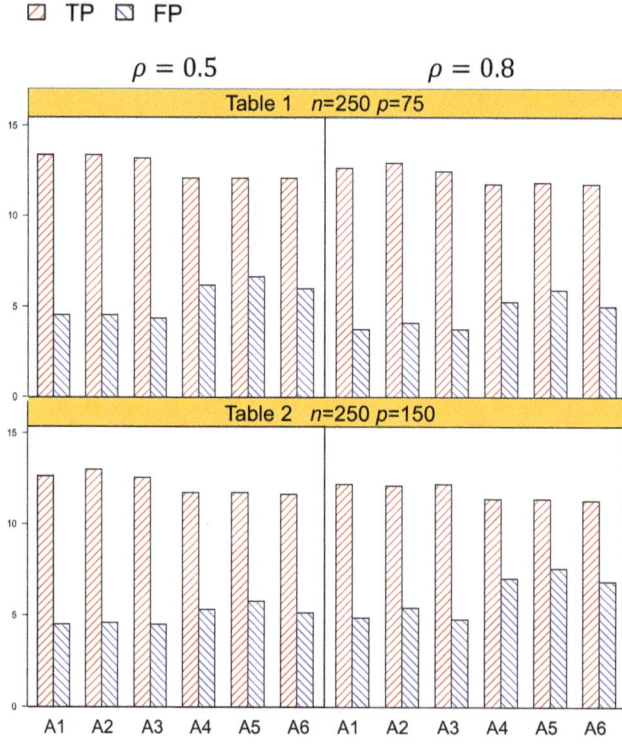

Figure 1. Plot of the identification results for $n = 250$, $p = 75$ with an actual dimension of 304. $p = 150$ with an actual dimension of 604. A1–A3: methods accommodating the lipid–environment interactions with exchangeable, AR(1), and independence working correlations, respectively. A4–A6: methods not accommodating the lipid–environment interactions with exchangeable, AR(1), and independence working correlations, respectively.

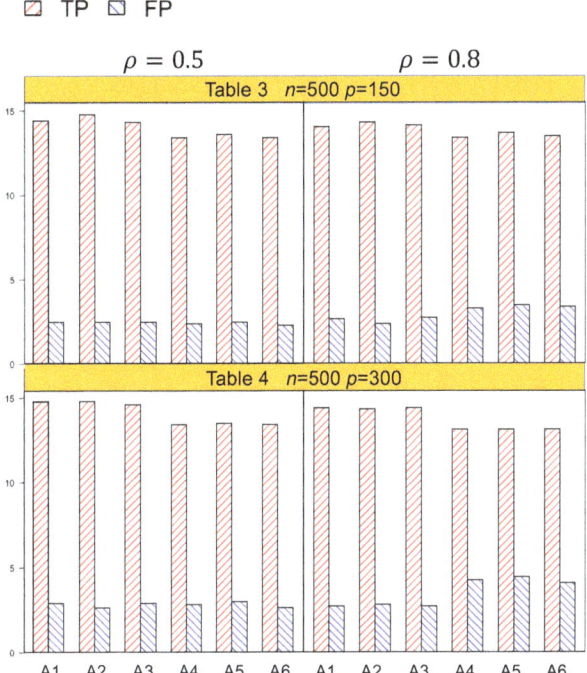

Figure 2. Plot of the identification results for $n = 500$, $p = 150$ with an actual dimension of 604. $p = 300$ with an actual dimension of 1204. A1–A3: methods accommodating the lipid–environment interactions with exchangeable, AR(1), and independence working correlations, respectively. A4–A6: methods not accommodating the lipid–environment interactions with exchangeable, AR(1), and independence working correlations, respectively.

In terms of estimation accuracy, A1–A3 also had a better performance compared with A4–A4, as shown in Tables 5 and 6. For the panel corresponding to $n = 250$, $\rho = 0.5$, and $p = 75$ in Table 5, the mean squared error for the nonzero coefficients of A1 was 0.1055, which was less than half of that of A4 (0.2321). Besides, A1 also had a smaller total mean squared error (TMSE). All the pieces of evidence suggested that A1 had higher estimation accuracy than A4. We can observe the pattern for the rest of the four methods. As the dimension increased to $n = 500$, $\rho = 0.8$, and $p = 300$ (so the total dimension was 1204) in Table 6, the MSE of A1 (0.0688) was also smaller than that of A4 (0.1949). There were no obvious differences in NMSE among these settings.

Another important conclusion we make from the simulation study is that, for the methods that differ only in working correlation, i.e., A1 (exchangeable), A2 (AR1), and A3 (independence), there was no significant difference in terms of either identification or estimation accuracy, as shown by Tables 1–6, as well as Figures 1 and 2. Such an observation suggests that the proposed methods under the GEE framework were robust to the misspecification of the working correlation, and this is consistent with the conclusions from main effects only models in longitudinal studies [7].

Table 5. Estimation accuracy results for $n = 250$, $p = 75$ with an actual dimension of 304. $p = 150$ with an actual dimension of 604.

		\multicolumn{6}{c}{$n = 250$}
		\multicolumn{3}{c}{$p = 75$}	\multicolumn{3}{c}{$p = 150$}
		MSE	NMSE	TMSE	MSE	NMSE	TMSE
$\rho = 0.5$	A1	0.1055	0.0026	0.0043	0.1264	0.0045	0.0072
	A2	0.1042	0.0026	0.0042	0.1259	0.0045	0.0072
	A3	0.1030	0.0026	0.0042	0.1174	0.0041	0.0066
	A4	0.2321	0.0018	0.0056	0.2435	0.0032	0.0084
	A5	0.2304	0.0018	0.0055	0.2402	0.0031	0.0082
	A6	0.2288	0.0018	0.0055	0.2346	0.0030	0.0080
$\rho = 0.8$	A1	0.1187	0.0087	0.0135	0.129	0.0048	0.0075
	A2	0.1163	0.0085	0.0132	0.1295	0.0048	0.0075
	A3	0.1066	0.0075	0.0118	0.1319	0.0049	0.0077
	A4	0.2410	0.0060	0.0162	0.2531	0.0038	0.0092
	A5	0.2426	0.0060	0.0162	0.2487	0.0038	0.0091
	A6	0.2335	0.0058	0.0157	0.2431	0.0037	0.0089

Mean (sd) based on 100 replicates. A1–A3: methods accommodating the lipid–environment interactions with exchangeable, AR(1), and independence working correlations, respectively. A4–A6: methods not accommodating the lipid–environment interactions with exchangeable, AR(1), and independence working correlations, respectively.

Table 6. Estimation accuracy results for $n = 500$, $p = 150$ with an actual dimension of 604. $p = 300$ with an actual dimension of 1204.

		\multicolumn{6}{c}{$n = 500$}
		\multicolumn{3}{c}{$p = 150$}	\multicolumn{3}{c}{$p = 300$}
		MSE	NMSE	TMSE	MSE	NMSE	TMSE
$\rho = 0.5$	A1	0.0754	0.0026	0.0042	0.0660	0.0010	0.0017
	A2	0.0731	0.0026	0.0041	0.0659	0.0010	0.0017
	A3	0.0648	0.0022	0.0035	0.0663	0.0010	0.0017
	A4	0.1872	0.0015	0.0055	0.1635	0.0007	0.0024
	A5	0.1837	0.0015	0.0054	0.1612	0.0007	0.0024
	A6	0.1792	0.0013	0.0052	0.1603	0.0007	0.0024
$\rho = 0.8$	A1	0.0708	0.0023	0.0037	0.0688	0.0010	0.0018
	A2	0.0716	0.0023	0.0038	0.0688	0.0011	0.0018
	A3	0.0704	0.0025	0.0039	0.0718	0.0012	0.0020
	A4	0.1480	0.0013	0.0049	0.1949	0.0007	0.0028
	A5	0.1492	0.0013	0.0045	0.1945	0.0007	0.0028
	A6	0.1479	0.0012	0.0044	0.1899	0.0007	0.0027

Mean (sd) based on 100 replicates. A1–A3: methods accommodating the lipid–environment interactions with exchangeable, AR(1), and independence working correlations, respectively. A4–A6: methods not accommodating the lipid–environment interactions with exchangeable, AR(1), and independence working correlations, respectively.

To mimic the sample size and number of lipid factors in the case study, we also conducted a simulation in settings with $n = 60$, $p = 30$, and $q = 3$. Therefore, the overall dimension of main and interaction effects was 124. The coefficients were generated from $U[1.4, 1.8]$ for 17 nonzero effects. The identification and prediction results are summarized in Tables A1 and A2 in the Appendix A, respectively. Consistent patterns were observed. For example, in terms of identification, under $\rho = 0.5$,

A1 had a higher TP of 13.6 (sd 2.5) compared to the 11.1 (sd 2.6) of A4, and a lower FP of 4.7 (sd 2.7), compared to the FP of 5.4 (sd 2.8) identified by A4.

Evaluation of all the methods, especially A1–A3, was also conducted when the true underlying model was misspecified. We generated the response (phenotype) from a main effect only model with eight true main effects when $n = 250$, $p = 75$, $\rho = 0.8$ with a total dimension of 304. Results are provided in Table A3. When the interaction effects did not exist, A1 had only identified a very small number of false interaction effects, with 0.7 (sd 1.7) false positives. A2–A6 performed similarly in terms of identifying false interaction effects. All six methods identified a comparable number of true main effects. Overall, all methods had similar performance in identification, as well as prediction, when the data generating model had only main effects. Such a phenomenon is reasonable by further examining the results in Table 1. We found that the major difference between A1–A3 and A4–A6 was due to the identification of interaction effects. Therefore, when only main effects were present, all the methods had comparable performances.

Penalized regression and hypothesis testing are two related, but distinct aspects in statistical analysis. The proposed study was not aimed at developing test statistics, computing the power functions, and assessing the control of type 1 error, so these statistical test related results are not available, just like most of the studies on penalized regression. Recently, efforts devoted to bridging the two areas have been mainly restricted to linear models under high-dimensional settings [24–26]. Extensions to interaction models have not been reported so far. In particular, we are not aware of results reported for longitudinal models. Nevertheless, we conducted the simulation by assuming the null model and tabulate the identification results in Table A4. The results should be interpreted as identification with misspecified models. As we observed, under the null model, all six methods led to a very small number of false positives.

To assess the consistency of variable selection in longitudinal settings, we carried out the stability selection [27] under $n = 250$, $p = 75$, and $\rho = 0.8$. Each time, we selected 200 out of the total of 250 subjects without replacement and then conducted selection. The process was repeated 100 times, which yielded a proportion of selected effects. Larger proportions of being selected suggested stable results. Stability selection is well known for assessing the stability of penalized selection, and it alleviates the concern that the effects have only been identified by chance. We investigate the selection proportions of the 17 true main and interaction effects for all six methods in Table A5. A1 identified 14 true effects with proportions above 70%, which is consistent with the results shown in the lower panel of Table 1, where 13.7 TPs (sd 2.3) were identified. Such a consistent pattern can be observed across all six methods.

Although no consensus on the optimal criterion of selecting tuning parameters has been reached so far, cross-validation is perhaps the most well accepted criterion to select tuning parameters in the community of high-dimensional data analysis [3,4]. To further justify its appropriateness, under the setting of $n = 250$ and $p = 75$, we performed the analysis by selecting tuning parameters using an independently generated testing dataset with a sample size of 1000 and $p = 75$. The models were fitted on the training dataset, and prediction was assessed based on the independently generated testing dataset, so no data were used in training the model. The identification and prediction results are tabulated in Tables A6 and A7, respectively. A comparison to Tables 1 and 5 demonstrates that the results obtained by cross-validation and validation were very close.

3.2. Real Data Analysis

We applied the proposed and alternative methods on a dataset from one of our previous studies in animal models [15]. In the study, 60 female CD-1 mice were assigned to four different treatment groups, which were control (ad libitum feeding and sedentary), AE (exercise and ad libitum feeding), PE (exercise and pair feeding), and DCR (sedentary and 20% dietary calorie restriction). The phenotype of interest was mice's body weight, which was measured every week for 10 weeks. Mice were sedentary and given ad libitum feeding in the control group, where they could eat as much as they wanted without doing treadmill exercises. In the AE group, mice received ad libitum feeding and ran on the treadmill every day at a speed of 0.5 mph, 1 hour per day, and 5 days a week, while mice in the PE group did the same exercise, but were given the same amount of diet as the mice in the control group. Mice in the DCR group had 20% less calorie intake than the control group, but they had the same intake of protein, vitamins, and minerals. The composition of 176 plasma neutral lipid species of interest was measured. In the current study, we only focused on diacylglycerols. In addition, the diacylglycerol lipid species that have a majority of samples lower than the detection limits were excluded so there were 31 diacylglycerols. In total, there were 31 lipid main effects and 93 lipid–environment interactions.

Using the method A1 (interep with the exchangeable working correlation) as shown in Table 7, we identified seven lipid species that had different effects in weight control of mice (AE, PE, or DCR) on body weight compared to those of the control mice. Among them, C20:1/16:1 and C20:1/20:4 had negative interactions in AE mice, where C denotes carbon. For the lipid species of C20:1/16:1, $C_{39}H_{76}O_5N$, the regression coefficient was -2.9145 for AE mice. That is, mice with an increased amount of C20:1/16:1 tended to have a lower body weight compared to that of the control. In the AE mice, both C16:0/C16:0 and C22:6/C18:1 had strong positive associations with body weights. It is interesting that C16:0/C16:0 were negatively associated with body weight in both PE and DCR mice. C16:0 is also called palmitic acid and is one of most common saturated fatty acids. Increased consumption of palmitic acid is associated with higher risk of cardiovascular disease, type 2 diabetes, and cancer [28]. The negative association of C16:0/16:0 and body weight in DCR and PE suggests that when the calories of the diet are restricted, the accumulation of saturated fat in the body actually decreased compared to the control. Another lipid that is negatively associated with body weight in DCR and PE mice is C18:1/16:1. The lipids that were positively associated with body weight in PE were C18:2/C16:1, C20:1/C16:1, and C22:6/C18:1. All species contain unsaturated fatty acids. Among them, C22:6 is one of the omega-3 polyunsaturated fatty acids (PUFA). In DCR, the two lipids that were positively associated with body weight were C18:2/16:1 and C20:1/20:4. Both fatty acids C18:2 and C20:4 were PUFA. The results seem to be consistent with our previous finding that exercise with paired feeding may increase the amount of PUFA in phospholipids in mice skin [29].

Table 7. Real data analysis result from method A1 (method accommodating the lipid–environment interactions with exchangeable working correlation).

	Lipid	AE	PE	DCR
C16:0/16:1	0	0.0117	-0.0239	-0.0057
C18:2/16:1	0	0.1544	3.3322	0.3924
C18:1/16:1	0	0.4857	-0.6299	-0.5559
C20:1/16:1	0.5966	-2.9145	0.1299	-1.4836
C16:0/16:0	0	1.3742	-0.8817	-1.8070
C20:6/16:0	0.0369	0	0	0
C20:0/18:3	-1.3628	0	0	0
C18:0/18:2	-1.6154	0	0	0
C22:6/18:1	1.1717	1.7526	0.2287	-0.4079
C18:2/20:4	1.1497	0	0	0
C18:1/20:4	0.8490	0	0	0
C20:1/20:4	0	-0.2169	-0.6096	3.0537

AE, exercise and ad libitum feeding; PE, exercise and pair feeding; DCR, sedentary and 20% dietary calorie restriction.

In addition, we adopted A4 to analyze the lipid data. A4 also had the exchangeable working correlation, but it could not conduct group level selection of the lipid–environment interactions. The identification results are tabulated in Table 8. Note that the selection of interactions with individual dummy environment factors was not consistent with the formulation of the lipid–environment interactions. In terms of prediction, A1 had a smaller prediction error (4.04) than that of A4 (4.97).

Table 8. Real data analysis result from method A4 (method not accommodating the lipid–environment interactions with exchangeable working correlation).

Lipid	AE	DCR	PE
C16:0/16:1	0	0	−0.0024	0
C18:2/16:1	−2.1856	0	3.2306	0
C18:1/16:1	0	0	−1.4641	−2.3563
C20:1/16:1	0.0042	−2.6768	0	−1.7757
C16:0/16:0	0	2.8757	−0.9389	−2.6791
C18:2/16:0	0	0	0	−1.7688
C20:6/16:0	0.1481	−0.1276	0	0
C18:1/18:3	0	0	1.2917	0
C20:0/18:3	−1.6171	0	0	0
C18:0/18:2	−1.7695	0	0	0
C22:6/18:1	0.8851	3.4714	0.4809	0
C18:1/18:0	0	−1.2901	0	0
C22:7/18:0	0	−0.9839	0	0
C18:2/20:4	2.5871	0.6150	0	1.9327
C18:1/20:4	0	0	−0.0031	0
C20:1/20:4	0.7542	−1.1147	0	3.5396

4. Discussion

Investigation of the potential roles of lipids in the regulation and control of cellular function and the interactions between lipids and environmental factors are very important in the understanding of physiology and disease processes. Traditionally, the analyses mostly focus on the total amount of a particular type of lipid, such as total triglyceride, total cholesterol, and omega-3 fatty acid. With the recent advances in instrumental technology, it is feasible to analyze quantitatively a broad range of lipid species in a single platform [13,15,30–32]. The vast arrays of data generated in lipid profiling studies bring challenges to the statistical analysis of lipidomics data [33–35].

In this study, we proposed a penalized variable selection method to identify important lipid–environmental effects in longitudinal studies. Some statistical methods have already been reported for lipidomics studies, including the marginal test and variable selection methods [15,32,34,35]; however, they cannot be directly extended to longitudinal studies. On the other hand, existing variable selection methods for longitudinal data have been predominately developed for the identification of main effects and cannot accommodate the group level interaction structure unique to our studies. Both the simulation and case study have convincingly demonstrated the merit of the proposed interep over alternatives.

We selected tuning parameters based on cross-validation. A further investigation of different tuning criteria is interesting, but beyond the scope of this study, especially given the fact that many well known variable selection methods in longitudinal studies, such as [5], have been conducted using cross-validation. To facilitate a fair cross-comparison with existing relevant studies, we believe it is reasonable to adopt cross-validation to choose tuning parameters. Note that the aforementioned stability selection analysis also partially justifies the usage of cross-validation. We acknowledge that other criteria for selecting tunings, such as double cross-validation [36], could be a potential reliable choice. However, as it is not a widely accepted tuning criterion for high-dimensional data analysis and has not been adopted in any longitudinal studies so far, we postpone the investigation to the future.

Interaction studies have been historically pursued by statisticians [37]. Within the high-dimensional scenario, accounting for such a complex structure, in both gene–gene (G × G) and gene–environment (G × E) interaction studies, is challenging, but also rewarding [38]. The proposed study is among the first to investigate penalized identification of lipid–environment interactions in longitudinal studies. Both the simulation study and case study yielded interesting findings. G × G interaction is computationally more challenging than G × E interactions since both main effects involved in the interactions are of high dimensionality. Following the representative G × G interaction studies [39,40], we can extend the proposed study to lipid–lipid interactions, which has not been investigated in longitudinal studies so far. Besides, when multi-omics measurements are available, it is also of great interest to examine interaction effects through multi-omics integration studies in the longitudinal setting [41,42].

The proposed model can also be estimated using the quadratic inference functions (QIF). GEE relies on the working correlation matrix $R(\eta)$, and it enables us to find the consistent estimator of the regression parameter if consistent estimators of the nuisance parameters η can be obtained. However, consistent estimators of η do not always exist in some cases. QIF has been proposed to avoid explicit estimation of the nuisance parameters by assuming the inverse of the working correlation matrix $R(\eta)$ can be approximated by a linear combination of a class of base matrices [7,43]. Thus, QIF is robust to the misspecification of the working correlation.

In this paper, we are interested in the identification of lipid-treatment (or environment) interactions through penalization. The success of set based analysis, including those for the gene set [44] and SNP set [45,46], has tremendously motivated the development of statistical methods for G × E interactions from marginal analyses ([47,48]) to penalization methods [17,18,49]. Our model can be potentially extended in the following aspects. First, as data contamination and outliers have been widely observed in repeated measurements, robust variable selection methods in G × E interaction studies [23,50–52] can be extended to longitudinal settings. Second, recently, multiple Bayesian methods have been proposed for pinpointing important G × E interaction effects [53–55]. Within the framework of analyzing repeated measurements, Bayesian variable selection for interactions has not been extensively examined. Investigations of all these possible directions will be postponed to the near future.

Author Contributions: Conceptualization, C.W. and Y.J.; resources, Y.J., W.W., and C.W.; methodology, F.Z., J.R., Y.J., and C.W.; writing, original draft preparation, F.Z. and C.W.; software, F.Z. and J.R.; data analysis, F.Z., Y.J., and C.W.; writing, review and editing, all authors; supervision, C.W. and Y.J.; project administration, C.W.; funding acquisition, C.W.

Funding: This study was partially supported by an Innovative Research Award from the Johnson Cancer Research Center at Kansas State University and a Kansas State University Faculty Enhancement Award.

Conflicts of Interest: The authors declare no conflict of interest.

Abbreviations

The following abbreviations are used in this manuscript:

GEE	Generalized estimating equation
AE	Exercise and ad libitum feeding
PE	Exercise and pair feeding
DCR	Sedentary and 20% dietary calorie restriction
TG	Triacylglycerol
DG	Diacylglycerol
LASSO	Least absolute shrinkage and selection operator
PGEE	Penalized generalized estimating equation
PQIF	Penalized quadratic inference function
MCP	Minimax concave penalty
SCAD	Smoothly clipped absolute deviation
SNP	Single nucleotide polymorphisms
CNV	Copy number variations
QIF	Quadratic inference function

Appendix A

Table A1. Identification results for $n = 60$, $p = 30$ with an actual dimension of 124.

$n = 60$	$p = 30$	Overall		Main		Interaction	
		TP	FP	TP	FP	TP	FP
$\rho = 0.5$	A1	13.6(2.5)	4.7(2.7)	7.4(0.8)	2.1(1.6)	6.2(2.1)	2.5(2.6)
	A2	13.6(2.5)	4.8(2.8)	7.3(0.8)	2.2(1.6)	6.2(2.1)	2.6(2.6)
	A3	13.7(2.5)	4.9(3.0)	7.4(0.7)	2.1(1.6)	6.3(2.1)	2.7(2.7)
	A4	11.1(2.6)	5.4(2.8)	6.4(1.1)	1.1(1.0)	4.6(1.9)	4.3(2.3)
	A5	11.1(2.6)	5.4(2.8)	6.4(1.1)	1.1(1.0)	4.6(1.9)	4.3(2.3)
	A6	11.1(2.5)	5.5(2.8)	6.5(1.2)	1.1(1.0)	4.7(1.8)	4.4(2.3)
$\rho = 0.8$	A1	13.2(2.2)	4.4(2.9)	7.5(0.6)	2.4(1.7)	5.7(2.1)	1.9(2.1)
	A2	13.2(2.2)	4.4(2.9)	7.5(0.6)	2.4(1.7)	5.7(2.1)	2.0(2.1)
	A3	13.4(2.0)	4.4(3.0)	7.5(0.6)	2.4(1.7)	5.9(1.9)	2.0(2.1)
	A4	11.0(2.4)	5.5(2.5)	6.5(1.4)	1.3(1.2)	4.5(1.8)	4.2(2.1)
	A5	11.0(2.4)	5.6(2.6)	6.5(1.4)	1.3(1.2)	4.5(1.8)	4.2(2.2)
	A6	11.1(2.4)	5.8(2.7)	6.5(1.4)	1.4(1.3)	4.5(1.8)	4.3(2.2)

Mean (sd) based on 100 replicates. A1–A3: methods accommodating the lipid–environment interactions with exchangeable, AR(1), and independence working correlations, respectively. A4–A6: methods not accommodating the lipid–environment interactions with exchangeable, AR(1), and independence working correlations, respectively.

Table A2. Estimation accuracy results for $n = 60$, $p = 30$ with an actual dimension of 124.

		$n = 60, p = 30$				
		$\rho = 0.5$			$\rho = 0.8$		
	MSE	NMSE	TMSE	MSE	NMSE	TMSE
A1	0.9352	0.1928	0.2732	0.9820	0.2108	0.2944
A2	0.9387	0.1924	0.2733	0.9809	0.2105	0.2940
A3	0.9324	0.1914	0.2717	1.0098	0.2063	0.2933
A4	1.9732	0.1560	0.3528	1.9910	0.1488	0.3484
A5	1.9709	0.1556	0.3523	1.9887	0.1487	0.348
A6	1.9629	0.1543	0.3502	1.9795	0.1474	0.3458

Mean (sd) based on 100 replicates. A1–A3: methods accommodating the lipid–environment interactions with exchangeable, AR(1), and independence working correlations, respectively. A4–A6: methods not accommodating the lipid–environment interactions with exchangeable, AR(1), and independence working correlations, respectively.

Table A3. Data simulated based on the underlying main effect only model. Identification results for $n = 250, p = 75, \rho = 0.8$ with an actual dimension of 304.

	Overall		Main		Interaction				
	TP	FP	TP	FP	TP	FP	MSE	NMSE	TMSE
A1	7.7(0.9)	0.7(1.7)	7.7(0.9)	0.0(0.0)	0.0(0.0)	0.7(1.7)	0.1025	0.0000	0.0014
A2	7.8(0.6)	0.4(1.3)	7.8(0.6)	0.0(0.2)	0.0(0.0)	0.4(1.3)	0.0730	0.0000	0.0010
A3	7.9(0.3)	0.5(1.2)	7.9(0.3)	0.3(0.7)	0.0(0.0)	0.2(0.8)	0.0288	0.0000	0.0004
A4	7.3(1.1)	0.8(0.9)	7.3(1.1)	0.0(0.0)	0.0(0.0)	0.8(0.9)	0.2530	0.0000	0.0034
A5	7.2(1.1)	0.9(1.1)	7.2(1.1)	0.0(0.0)	0.0(0.0)	0.9(1.1)	0.2273	0.0001	0.0031
A6	7.5(0.7)	1.2(1.1)	7.5(0.7)	0.0(0.2)	0.0(0.0)	1.2(1.1)	0.1932	0.0001	0.0027

Mean (sd) based on 100 replicates. A1–A3: methods accommodating the lipid–environment interactions with exchangeable, AR(1), and independence working correlations, respectively. A4–A6: methods not accommodating the lipid–environment interactions with exchangeable, AR(1), and independence working correlations, respectively.

Table A4. Null models.

	$n = 250$				$n = 500$			
	$p = 75$		$p = 150$		$p = 150$		$p = 300$	
	$\rho = 0.5$	$\rho = 0.8$	$\rho = 0.5$	$\rho = 0.8$	$\rho = 0.5$	$\rho = 0.8$	$\rho = 0.5$	$\rho = 0.8$
A1	0.00(0.00)	0.03(0.18)	0.03(0.18)	0.00(0.00)	0.00(0.00)	0.00(0.00)	0.00(0.00)	0.00(0.00)
A2	0.03(0.10)	0.03(0.18)	0.30(0.70)	0.10(0.31)	0.00(0.00)	0.00(0.00)	0.03(0.18)	0.00(0.00)
A3	0.13(0.51)	0.17(0.44)	0.97(1.47)	0.77(0.81)	0.10(0.40)	0.50(0.20)	0.10(0.31)	0.10(0.25)
A4	0.00(0.00)	0.03(0.18)	0.03(0.18)	0.00(0.00)	0.00(0.00)	0.00(0.00)	0.00(0.00)	0.00(0.00)
A5	0.03(0.10)	0.03(0.18)	0.30(0.70)	0.10(0.31)	0.00(0.00)	0.00(0.00)	0.03(0.18)	0.00(0.00)
A6	0.13(0.51)	0.17(0.44)	0.97(1.47)	0.77(0.81)	0.10(0.40)	0.50(0.20)	0.10(0.31)	0.10(0.25)

Mean (sd) based on 100 replicates. A1–A3: methods accommodating the lipid–environment interactions with exchangeable, AR(1), and independence working correlations, respectively. A4–A6: methods not accommodating the lipid–environment interactions with exchangeable, AR(1), and independence working correlations, respectively.

Table A5. Stability selection percentages for all 17 true effects in the simulated data when $n = 250$, $p = 75, \rho = 0.8$ with an actual dimension of 304.

True Effect	A1	A2	A3	A4	A5	A6
1	1	1	1	1	1	1
2	0.73	1	1	0.82	0.98	1
3	1	0.80	1	1	1	1
4	1	1	1	1	1	1
5	1	0.45	1	1	0.93	0.98
6	0.13	0.14	0.38	0.65	0.98	0.98
7	0.58	0.65	1	0.99	1	0.92
8	0.61	0.25	0.45	0.89	1	1
9	1	0.84	1	0.46	0.02	0.10
10	1	0.86	1	0.07	0.01	0.10
11	1	0.83	1	0.7	0.66	0.84
12	0.77	0.91	0.72	0.36	0.87	0.01
13	0.77	0.91	0.73	0.39	0.94	0.45
14	0.75	0.94	0.77	0.48	1	0.98
15	0.81	0.82	0.98	0.30	0.55	1
16	0.80	0.86	0.99	0.98	0.75	0.99
17	0.80	0.87	0.99	0.66	0.93	1

A1–A3: methods accommodating the lipid–environment interactions with exchangeable, AR(1), and independence working correlations, respectively. A4–A6: methods not accommodating the lipid–environment interactions with exchangeable, AR(1), and independence working correlations, respectively.

Table A6. Validation methods. Identification results for $n = 250$, $p = 75$ with an actual dimension of 304.

$n = 250$	$p = 75$	Overall		Main		Interaction	
		TP	FP	TP	FP	TP	FP
$\rho=0.5$	A1	14.1(2.1)	4.6(3.1)	7.0(0.8)	1.1(0.8)	7.0(1.8)	3.5(2.9)
	A2	14.2(2.1)	4.7(3.1)	7.0(0.9)	1.1(0.9)	7.1(1.8)	3.6(2.8)
	A3	14.4(1.7)	4.6(3.2)	7.1(0.8)	1.1(0.9)	7.2(1.5)	3.5(3.0)
	A4	13.1(1.1)	6.1(2.8)	6.9(0.8)	1.0(0.8)	6.1(0.9)	5.3(2.6)
	A5	13.1(1.1)	6.4(2.8)	6.9(0.8)	1.0(0.8)	6.1(0.9)	5.6(2.5)
	A6	13.0(1.2)	6.7(3.1)	6.9(0.8)	1.0(0.8)	6.1(1.0)	5.9(2.9)
$\rho=0.8$	A1	13.7(2.6)	4.7(2.9)	7.2(0.8)	1.4(0.9)	6.5(2.3)	3.2(2.5)
	A2	13.8(2.6)	4.6(3.1)	7.3(0.8)	1.4(1.0)	6.6(2.3)	3.1(2.6)
	A3	13.8(2.5)	5.1(3.0)	7.3(0.7)	1.5(0.8)	6.5(2.1)	3.6(2.9)
	A4	12.9(2.1)	5.7(2.5)	7.3(0.8)	1.3(0.9)	5.6(1.6)	4.5(2.1)
	A5	12.9(2.1)	5.8(2.6)	7.3(0.8)	1.3(1.0)	5.6(1.6)	4.5(2.2)
	A6	12.9(2.2)	6.8(2.7)	7.3(0.7)	1.4(0.9)	5.6(1.8)	5.5(2.5)

Mean (sd) based on 100 replicates. A1–A3: methods accommodating the lipid–environment interactions with exchangeable, AR(1), and independence working correlations, respectively. A4–A6: methods not accommodating the lipid–environment interactions with exchangeable, AR(1), and independence working correlations, respectively.

Table A7. Validation methods. Estimation accuracy results for $n = 250$, $p = 75$ with an actual dimension of 304.

	$n = 250, p = 75$					
	$\rho = 0.5$			$\rho = 0.8$		
	MSE	NMSE	TMSE	MSE	NMSE	TMSE
A1	0.1126	0.0074	0.0120	0.1205	0.0085	0.0134
A2	0.1095	0.0071	0.0115	0.1200	0.0085	0.0133
A3	0.1082	0.0071	0.0115	0.1245	0.0090	0.0140
A4	0.2344	0.0051	0.0150	0.2610	0.0060	0.0171
A5	0.2335	0.0050	0.0149	0.2627	0.0060	0.0171
A6	0.2302	0.0048	0.0146	0.2565	0.0058	0.0166

Mean (sd) based on 100 replicates. A1–A3: methods accommodating the lipid–environment interactions with exchangeable, AR(1), and independence working correlations, respectively. A4–A6: methods not accommodating the lipid–environment interactions with exchangeable, AR(1), and independence working correlations, respectively.

References

1. Verbeke, G.; Fieuws, S.; Molenberghs, G.; Davidian, M. The analysis of multivariate longitudinal data: A review. *Stat. Methods Med. Res.* **2014**, *23*, 42–59 [CrossRef] [PubMed]
2. Bandyopadhyay, S.; Ganguli, B.; Chatterjee, A. A review of multivariate longitudinal data analysis. *Stat. Methods Med. Res.* **2011**, *20*, 299–330. [CrossRef] [PubMed]
3. Fan, J.; Lv, J. A selective overview of variable selection in high-dimensional feature space. *Stat. Sin.* **2010**, *20*, 101–148. [PubMed]
4. Wu, C.; Ma, S. A selective review of robust variable selection with applications in bioinformatics. *Brief. Bioinform.* **2014**, *16*, 873–883. [CrossRef] [PubMed]
5. Wang, L.; Zhou, J.; Qu, A. Penalized generalized estimating equations for high-dimensional longitudinal data analysis. *Biometrics* **2012**, *68*, 353–360. [CrossRef]
6. Ma, S.; Song, Q.; Wang, L. Simultaneous variable selection and estimation in semiparametric modeling of longitudinal/clustered data. *Bernoulli* **2013**, *19*, 252–274. [CrossRef]
7. Cho, H.; Qu, A. Model selection for correlated data with diverging number of parameters. *Stat. Sin.* **2013**, *23*, 901–927. [CrossRef]

8. Berridge, M.J. Inositol trisphosphate and diacylglycerol: Two interacting second messengers. *Annu. Rev. Biochem.* **1987**, *56*, 159–193. [CrossRef]
9. Goñi, F.M.; Alonso, A. Structure and functional properties of diacylglycerols in membranes. *Prog. Lipid Res.* **1999**, *38*, 1–48. [CrossRef]
10. Barona, T.; Byrne, R.D.; Pettitt, T.R.; Wakelam, M.J.; Larijani, B.; Poccia, D.L. Diacylglycerol induces fusion of nuclear envelope membrane precursor vesicles. *J. Biol. Chem.* **2005**, *280*, 41171–41177. [CrossRef]
11. Thiam, A.R.; Farese, R.V., Jr.; Walther, T.C. The biophysics and cell biology of lipid droplets. *Nat. Rev. Mol. Cell Biol.* **2013**, *14*, 775–786. [CrossRef] [PubMed]
12. Markgraf, D.; Al-Hasani, H.; Lehr, S. Lipidomics—Reshaping the analysis and perception of type 2 diabetes. *Int. J. Mol. Sci.* **2016**, *17*, 1841. [CrossRef] [PubMed]
13. Zhou, X.; Mao, J.; Ai, J.; Deng, Y.; Roth, M.R.; Pound, C.; Henegar, J.; Welti, R.; Bigler, S.A. Identification of plasma lipid biomarkers for prostate cancer by lipidomics and bioinformatics. *PLoS ONE* **2012**, *7*, e48889. [CrossRef] [PubMed]
14. Stephenson, D.J.; Hoeferlin, L.A.; Chalfant, C.E. Lipidomics in translational research and the clinical significance of lipid–based biomarkers. *Transl. Res.* **2017**, *189*, 13–29. [CrossRef] [PubMed]
15. King, B.S.; Lu, L.; Yu, M.; Jiang, Y.; Standard, J.; Su, X.; Zhao, Z.; Wang, W. Lipidomic profiling of di–and tri–acylglycerol species in weight-controlled mice. *PLoS ONE* **2015**, *10*, e0116398. [CrossRef]
16. Yuan, M.; Lin, Y. Model selection and estimation in regression with grouped variables. *J. R. Stat. Soc. Ser. (Stat. Methodol.)* **2006**, *68*, 49–67. [CrossRef]
17. Wu, C.; Cui, Y.; Ma, S. Integrative analysis of gene–environment interactions under a multi-response partially linear varying coefficient model. *Stat. Med.* **2014**, *33*, 4988–4998. [CrossRef]
18. Wu, C.; Zhong, P.S.; Cui, Y. Additive varying-coefficient model for nonlinear gene-environment interactions. *Stat. Appl. Genet. Mol. Biol.* **2018**, *17*. [CrossRef]
19. Fan, Y.; Qin, G.; Zhu, Z. Variable selection in robust regression models for longitudinal data. *J. Multivar. Anal.* **2012**, *109*, 156–167. [CrossRef]
20. Liang, K.Y.; Zeger, S.L. Longitudinal data analysis using generalized linear models. *Biometrika* **1986**, *73*, 13–22. [CrossRef]
21. Zhou, F.; Ren, J.; Li, X.; Wu, C.; Jiang, Y. *Interep: Interaction Analysis of Repeated Measure Data*; Version 0.3.0; 2019. Available online: https://rdrr.io/cran/interep/ (accessed on 26 November 2019).
22. Zhang, C.H. Nearly unbiased variable selection under minimax concave penalty. *Ann. Stat.* **2010**, *38*, 894–942. [CrossRef]
23. Wu, C.; Jiang, Y.; Ren, J.; Cui, Y.; Ma, S. Dissecting gene–environment interactions: A penalized robust approach accounting for hierarchical structures. *Stat. Med.* **2018**, *37*, 437–456. [CrossRef] [PubMed]
24. Lockhart, R.; Taylor, J.; Tibshirani, R.J.; Tibshirani, R. A significance test for the lasso. *Ann. Stat.* **2014**, *42*, 413–468. [CrossRef] [PubMed]
25. Taylor, J.; Tibshirani, R.J. Statistical learning and selective inference. *Proc. Natl. Acad. Sci. USA* **2015**, *112*, 7629–7634. [CrossRef] [PubMed]
26. Lee, J.D.; Sun, D.L.; Sun, Y.; Taylor, J.E. Exact post-selection inference, with application to the lasso. *Ann. Stat.* **2016**, *44*, 907–927. [CrossRef]
27. Meinshausen, N.; Bühlmann, P. Stability selection. *J. R. Stat. Soc. Ser. (Stat. Methodol.)* **2010**, *72*, 417–473. [CrossRef]
28. Briggs, M.; Petersen, K.; Kris-Etherton, P. Saturated fatty acids and cardiovascular disease: Replacements for saturated fat to reduce cardiovascular risk. *Healthcare* **2017**, *5*, 29. [CrossRef]
29. Ouyang, P.; Jiang, Y.; Doan, H.M.; Xie, L.; Vasquez, D.; Welti, R.; Su, X.; Lu, N.; Herndon, B.; Yang, S.; et al. Weight Loss via exercise with controlled dietary intake may affect phospholipid profile for cancer prevention in murine skin tissues. *Cancer Prev. Res.* **2010**, *3*, 466–477. [CrossRef]
30. Bowden, J.A.; Heckert, A.; Ulmer, C.Z.; Jones, C.M.; Koelmel, J.P.; Abdullah, L.; Ahonen, L.; Alnouti, Y.; Armando, A.; Asara, J.M.; et al. Harmonizing lipidomics: NIST interlaboratory comparison exercise for lipidomics using standard reference material 1950 metabolites in frozen human plasma. *J. Lipid Res.* **2017**. [CrossRef]
31. Stegemann, C.; Pechlaner, R.; Willeit, P.; Langley, S. R.; Mangino, M.; Mayr, U.; Menni, C.; Moayyeri, A.; Santer, P.; Rungger, G.; et al. Lipidomics profiling and risk of cardiovascular disease in the prospective population-based Bruneck study. *Circulation* **2014**, *129*, 1821–1831. [CrossRef]

32. Jiang, Y.; Ma, H.; Su, X.; Chen, J.; Xu, J.; Standard, J.; Lin, D.; Wang, W. IGF-1 mediates exercise-induced phospholipid alteration in the murine skin tissues. *J. Nutr. Food Sci.* **2012**, *2*, 1–6. [CrossRef]
33. Wenk, M.R. The emerging field of lipidomics. *Nat. Rev. Drug Discov.* **2005**, *4*, 594. [CrossRef] [PubMed]
34. Kujala, M.; Nevalainen, J. A case study of normalization, missing data and variable selection methods in lipidomics. *Stat. Med.* **2015**, *34*, 59–73. [CrossRef]
35. Checa, A.; Bedia, C.; Jaumot, J. Lipidomic data analysis: Tutorial, practical guidelines and applications. *Anal. Chim. Acta* **2015**, *885*, 1–16. [CrossRef] [PubMed]
36. Filzmoser, P.; Liebmann, B.; Varmuza, K. Repeated double cross validation. *J. Chemom. J. Chemom. Soc.* **2009**, *23*, 160–171. [CrossRef]
37. Cordell, H.J. Epistasis: What it means, what it doesn't mean, and statistical methods to detect it in humans. *Hum. Mol. Genet.* **2002**, *11*, 2463–2468. [CrossRef]
38. Wu, M.; Ma, S. Robust genetic interaction analysis. *Brief. Bioinform.* **2018**, *20*, 624–637. [CrossRef]
39. Choi, N. H.; Li, W.; Zhu, J. Variable selection with the strong heredity constraint and its oracle property. *J. Am. Stat. Assoc.* **2010**, *105*, 354–364. [CrossRef]
40. Bien, J.; Taylor, J.; Tibshirani, R. A lasso for hierarchical interactions. *Ann. Stat.* **2013**, *41*, 1111–1141. [CrossRef]
41. Li, J.; Lu, Q.; Wen, Y. Multi-kernel linear mixed model with adaptive lasso for prediction analysis on high-dimensional multi-omics data. *Bioinformatics* **2019**, 1–10, in press.
42. Wu, C.; Zhou, F.; Ren, J.; Li, X.; Jiang, Y.; Ma, S. A selective review of multi-level omics data integration using variable selection. *High-Throughput* **2019**, *8*, 4. [CrossRef]
43. Qu, A.; Lindsay, B.G.; Li, B. Improving generalised estimating equations using quadratic inference functions. *Biometrika* **2000**, *87*, 823–836. [CrossRef]
44. Schaid, D.J.; Sinnwell, J.P.; Jenkins, G.D.; McDonnell, S.K.; Ingle, J.N.; Kubo, M.; Goss, P.E.; Costantino, J.P.; Wickerham, D.L.; Weinshilboum, R.M. Using the gene ontology to scan multilevel gene sets for associations in genome wide association studies. *Genet. Epidemiol.* **2012**, *36*, 3–16. [CrossRef] [PubMed]
45. Wu, C.; Cui, Y. Boosting signals in gene–based association studies via efficient SNP selection. *Brief. Bioinform.* **2013**, *15*, 279–291. [CrossRef] [PubMed]
46. Wu, C.; Li, S.; Cui, Y. Genetic association studies: An information content perspective. *Curr. Genom.* **2012**, *13*, 566–573. [CrossRef] [PubMed]
47. Mukherjee, B.; Ahn, J.; Gruber, S. B.; Chatterjee, N. Testing gene–environment interaction in large-scale case-control association studies: Possible choices and comparisons. *Am. J. Epidemiol.* **2011**, *175*, 177–190. [CrossRef] [PubMed]
48. Wu, C.; Cui, Y. A novel method for identifying nonlinear gene–environment interactions in case–control association studies. *Hum. Genet.* **2013**, *132*, 1413–1425 [CrossRef]
49. Wu, M.; Zhang, Q.; Ma, S. Structured gene–environment interaction analysis. *Biometrics* **2019**, 1–13, in press. [CrossRef]
50. Xu, Y.; Wu, M.; Ma, S.; Ejaz Ahmed, S. Robust gene–environment interaction analysis using penalized trimmed regression. *J. Stat. Comput. Simul.* **2018**, *88*, 3502–3528. [CrossRef]
51. Wu, C.; Shi, X.; Cui, Y.; Ma, S. A penalized robust semiparametric approach for gene–environment interactions. *Stat. Med.* **2015**, *34*, 4016–4030. [CrossRef]
52. Wu, M.; Ma, S. Robust semiparametric gene–environment interaction analysis using sparse boosting. *Stat. Med.* **2019**, in press. [CrossRef] [PubMed]
53. Ren, J.; Zhou, F.; Li, X.; Chen, Q.; Zhang, H.; Ma, S.; Jiang, Y.; Wu, C. Semi-parametric Bayesian variable selection for gene–environment interactions. *Stat. Med.* **2019**, 1–51, in press.
54. Li, J.; Wang, Z.; Li, R.; Wu, R. Bayesian group LASSO for nonparametric varying-coefficient models with application to functional genome–wide association studies. *Ann. Appl. Stat.* **2015**, *9*, 640–664. [CrossRef] [PubMed]
55. Ahn, J.; Mukherjee, B.; Gruber, S.B.; Ghosh, M. Bayesian semiparametric analysis for two-phase studies of gene–environment interaction. *Ann. Appl. Stat.* **2013**, *7*, 543–569. [CrossRef] [PubMed]

© 2019 by the authors. Licensee MDPI, Basel, Switzerland. This article is an open access article distributed under the terms and conditions of the Creative Commons Attribution (CC BY) license (http://creativecommons.org/licenses/by/4.0/).

MDPI
St. Alban-Anlage 66
4052 Basel
Switzerland
Tel. +41 61 683 77 34
Fax +41 61 302 89 18
www.mdpi.com

Genes Editorial Office
E-mail: genes@mdpi.com
www.mdpi.com/journal/genes

www.ingramcontent.com/pod-product-compliance
Lightning Source LLC
LaVergne TN
LVHW070559100526
838202LV00012B/506